P. Schattschneider

Fundamentals of Inelastic Electron Scattering

Springer-Verlag Wien New York

Dipl.-Ing. Dr. Peter Schattschneider
Institut für Angewandte und Technische Physik
Technische Universität Wien, Austria

With 74 Figures

Library of Congress Cataloging-in-Publication Data

Schattschneider, P. (Peter), 1950–
    Fundamentals of inelastic electron scattering.

    Bibliography: p.
    Includes indexes.
    1. Electrons--Scattering. I. Title.
QC793.5.E628S33 1986   539.7'54   86–17803

ISBN-13:978-3-211-81937-1   e-ISBN-13:978-3-7091-8866-8
DOI: 10.1007/978-3-7091-8866-8

To the memory of Roman Sexl

Foreword

Electron energy loss spectroscopy (ELS) is a vast subject with a long and honorable history. The problem of stopping power for high energy particles interested the earliest pioneers of quantum mechanics such as Bohr and Bethe, who laid the theoretical foundations of the subject. The experimental origins might perhaps be traced to the original Franck-Hertz experiment. The modern field includes topics as diverse as low energy reflection electron energy loss studies of surface vibrational modes, the spectroscopy of gases and the modern theory of plasmon excitation in crystals. For the study of ELS in electron microscopy, several historically distinct areas of physics are relevant, including the theory of the Debye-Waller factor for virtual inelastic scattering, the use of complex optical potentials, lattice dynamics for crystalline specimens and the theory of atomic ionisation for isolated atoms.

However the field of electron energy loss spectroscopy contains few useful texts which can be recommended for students. With the recent appearance of Raether's and Egerton's books (see text for references), we have for the first time both a comprehensive review text—due to Raether—and a lucid introductory text which emphasizes experimental aspects—due to Egerton. Raether's text tends to emphasize the recent work on surface plasmons, while the strength of Egerton's book is its treatment of inner shell excitations for microanalysis, based on the use of atomic wavefunctions for crystal electrons.

The present volume provides a different emphasis through the extensive use of the dielectric formulation, within which all the many mechanisms of ELS may be described. A general treatment of ELS has yet to appear, and may never do so, since this would be a vast undertaking. Such a treatment must allow for diffraction and energy loss processes of both the excited crystal electron and the beam electron for electronic excitation, while giving, in addition, a complete account of phonon losses (which require, if modes are to be included, a knowledge of lattice dynamics), together with the theory of plasmon excitations. Approaches to all these problems have appeared in the literature from various points of view, notably in the classic papers of Y. Kainuma, the first to give a general theory of combined multiple elastic Bragg scattering and electronic excitation—see reference [8.27]—, of A. Howie [8.28] for plasmon and elastic Bragg scattering, and of S. Takagi [8.29] for phonons and elastic Bragg scattering. More recently, from closely related synchrotron X-ray absorption studies on crystals we have come to learn how the near and extended absorption edge fine structure is related to the crystal density of states and electronic band structure. From the equivalent multiple scattering view-point, we have seen how the diffraction of an ejected core electron controls the inner shell edge shapes in ELS (see, amongst many, for example, Disko *et al.* [3.27]).

This book takes a fresh point of view, and provides, I believe, a valuable introduction to the theory of ELS by emphasizing the dielectric formulation, the form factor, Green functions, second quantisation and Feynman diagrams. This powerful approach is not used extensively in other ELS books, and the pedagocically sound and historically rich introduction to this description of ELS will be greatly valued by graduate students in many countries. Readers wishing to continue their study of this approach may wish to consult the excellent texts by S. Raimes and Y. Ohtsuki referenced in the text. The four papers mentioned above should also be consulted for a balanced view of the field.

Tempe, 1986   J. C. H. Spence

Preface

Curiously, a preface is the very last part of a manuscript to be written, but there is a good reason to use it as a frontispiece: It is wise to apologize for having omitted some important material that everybody knows about—except for the author.

This preface differs from the typical preface. To begin with, I do not apologize for the disregard of a number of items. For instance, the important issue of magnetic interactions between electrons and matter is completely ignored in the text. On the other hand, electron scattering is discussed at length in the framework of electrodynamics, in an effort to reconcile the phenomenological description of dielectrics with scattering mechanisms.

The book is arranged in such a way as to cover three of the most important aspects of inelastic electron scattering: interactions with atoms; dielectric response to a probe electron; and the quantum mechanical description of scattering by means of many-body theory. In the latter two parts, emphasis is put on collective phenomena (occurrence of plasmons).

The reader is warned against using the present volume as a mere textbook on interactions of electrons and matter—there are more comprehensive ones. It is not intended as a monograph either—there are more detailed ones covering any of the chapters. However, there is no satisfactory presentation of important and basic issues of inelastic scattering theory available at present. My intention was to deliver—and my hope is having written a treatise on fundamental questions encountered in inelastic electron scattering. The text supplies a rationale for understanding scattering phenomena; the theoretical background for the most important experimental methods; and—hopefully— an incentive to go into further details of the theory.

The present text emerged from a series of lectures held at the Technical University Vienna, Austria and at Changchun University, China. I am indebted to all of the students who detected more errors than there are formulae in the text. I have also extremely benefited from discussions with P. Kasperkovits, P. Pongratz and G. Sölkner. R.F. Egerton, D. Grau, J.C.H. Spence, P. Varga and

A. Wagendristel took great pain to read parts of the manuscript—many thanks to all of them for valuable suggestions.

I wish to thank K. Poppenberger for writing and assembling the manuscript and for her patience in the countless proof stages. I am grateful to O. Oberhauser for skilful advice and help in literature retrieval. F. Födermayr, D. Fritscher and K. Riedl helped a great deal in computation and word processing. The support of the EDV-Zentrum, Abt. Digitalrechenanlage at the Technical University Vienna and a grant of the Austrian Fonds zur Förderung der wissenschaftlichen Forschung which made the emergence of parts of the last chapter possible are gratefully acknowledged.

Thanks are also due to E. Korner, W. Koblitz and A. Kuzmann for early drafts of the manuscript, to L. Hametner and A. Kerper for assistance in preparing the figures, and to M. Fiebinger for improving my English.

Finally, I owe a great deal to P. Skalicky for his steadfast encouragement. My wife indulged in my temporary overestimation of the importance of electrons, and so contributed much to the creation of this book.

Vienna, May 1986 P. Schattschneider

Outline

Inelastic interactions of electrons with matter play an important role in many physical aspects. A familiar example is the excitation of atoms by transfer of a characteristic amount of kinetic energy from a fast impinging electron, a process utilized for generation of X-rays when the excited atom returns to the ground state. But also arc discharges and even ordinary electric lamps are based on transfer of energy from electrons to matter. As a more dramatic example, we mention the efforts to control deuterium fusion where scattering processes between ions and electrons in the plasma are an outstanding problem.

The notation *plasma* for a collective of mobile charged particles was first used by Langmuir in 1926. It soon turned out that a variety of unexpected phenomena occur in plasmas, and both experimentalists and theorists discovered it as a new field of work. Even nowadays it is far from being harvested. This is due to two facts: firstly, the theory of "the plasma", including all its manifestations from the dilute plasmas in interplanetary space with less than 10^9 charge carriers/m^3 to the very dense, degenerate solid state plasmas which are formed by the mobile conduction electrons in metals (about 10^{30} carriers/m^3), has not yet been solved generally. So far, a complete theory for the weakly coupled plasma is at hand, i.e. a plasma in which the interaction between the charge carriers is "small" in some sense. Secondly, solid state plasmas provide a rather new item for the experimentalist.

The first observation of a plasma phenomenon in the solid state was made in 1941 by Ruthemann when he found a strong oscillation of conduction electrons in thin Al films. At the time he did not realize what he had discovered. Experimental devices have been improved ever since; though only in the last two decades have experiments gained enough refinement and have been performed so extensively that results may merge with theory.

Electrons are an ideal probe for plasmas, especially in the solid state, because of their strong interaction with charged particles and small de Broglie-wavelength. The latter allows for high spatial resolution in the experiment whereas the strong Coulomb interaction with the constituents of the plasma leads to a considerable probability for transfer of energy and momentum between probe and specimen, i.e. to an inelastic scattering process of the probe electrons which can easily be monitored using electron optics.

This strong Coulomb interaction between charged particles suggests using electrons as a suitable probe of atomic distances, not only for plasmas since all matter is built of interacting charges on that scale. Electrons have been used for investigation of matter since their discovery in 1897 by J. J. Thomson. Soon it was realized that the scattering cross section of matter for electrons decreases with increasing velocity—the faster the electron, the more matter can it transmit. Classical models, assuming pointlike particles scattered by a Coulomb potential could explain the early experimental results. In 1921, when Ramsauer and Townsend independently investigated argon by use of slow electrons, they found startling evidence that the gas was nearly transparent for electrons of a particular small kinetic energy. Classical physics could not explain this observation.

Only quantum mechanics lateron afforded an explanation of the Ramsauer-Townsend effect in terms of phase shifts of the incident electron wave function. In the sequel atomic physics concentrated on the prediction of scattering cross sections of atoms. A wave mechanical formulation of the problem was originated in 1930 by H. Bethe. Today's calculations are, apart from many refinements, still founded on Bethe's theory. However, it can be said that there is neither a complete, general theory of inelastic collisions of electrons with atoms covering all experimental results, nor are there sufficiently accurate and numerous measurements to test for the validity of the various existing theoretical approaches.

The state of affairs in atomic physics was paralleled by the investigation of the solid state. In 1927 Davisson and Germer directly proved the wave properties of electrons when they obtained diffraction patterns from fast electrons transmitted through Zn foils. The stronger interaction of electrons with matter and their small de Broglie-wavelength make them superior to X-ray diffraction and soon led to the design of the electron microscope, the first of which was commercially available in 1936. In electron microscopy and electron diffraction, elastic collisions of the probe with matter are utilized. These techniques, developed much earlier than methods involving inelastic scattering, yield information on the static structure of matter (such as lattice type and lattice potential) which is described by the static structure factor $S(k)$. Inelastic scattering is mostly considered an undesired side effect in electron microscopy. Images get blurred, the strictly periodic intensity oscillations of the diffraction beam known as thickness fringes, and rocking curves get damped by inelastic processes. Electron diffractionists have tried more or less successfully to include these processes into dynamical diffraction theory. At present, a number of approximations are available, but there is no complete theory wherein inelastic interactions are treated at the same level of complexity as elastic ones.

Inelastic collision experiments may yield far more information on the specimen than elastic ones. This is evidenced when one considers the variety of inelastic "channels" (energy losses) which are open to the probe electron (such as phonon losses, electronic interband transitions, excitation of Cerenkov radiation) as compared to the single elastic channel utilized in electron microscopy and diffraction. Inelastic scattering experiments yield, contrary to the static case, the dynamical structure factor $S(k, \omega)$ which contains all the *dynamic* properties of the specimen probed. In passing we note that Synchrotron radiation is a powerful and increasingly important alternative to determine $S(k, \omega)$. - Most TEMs are nowadays equipped with such devices as microprobes or secondary electron detectors which utilize inelastic interactions.

Instead of using X-rays, secondary electrons or Auger electrons, the energy of the *primary* electron can be determined so as to monitor the inelastic interaction, hence energy losses which it had suffered. Although energy loss spectroscopy (ELS) exhibits a number of advantages, instruments and techniques have been developed only in the last two decades. This is mainly due to the relative complexity of spectrometers and to the initial difficulty in interpreting spectra.

At present ELS has evolved to a respectable tool in solid state physics, both for analytical and scientific purposes. Absorption edges in the spectra (which are the ELS counterparts of X-ray absorption edges) are used for quantitative microanalysis. They dominate the high energy region, whereas in the low energy loss region (up to ~ 20 eV) interband transitions are predominant. The latter are interpreted in terms of the complex dielectric function, similar to optical methods. Information on the electronic properties of the probed samples can be drawn from these parts of spectra. Both interband transitions and absorption edges are found in optical spectra, too. A genuine feature in ELS, not shared by any optical spectroscopy technique, is the occurrence of collective oscillations (plasmons) at ~ 10 eV–20 eV. These are longitudinal coherent oscillations of the loosely bound valence electrons which do not couple directly to transverse optical probes, but couple strongly to electrons. Plasmon spectra offer a new source of information on the solid state; however, their interpretation is by no means straightforward.

Reflection techniques with pre-specimen monochromators are commonly used. Despite their extremely high resolution (~ 10 meV) they suffer from interacting mainly with the surface of the specimen due to low penetration depths. On the other hand, in transmission spectroscopy of fast electrons (TEELS of FEELS) at primary energies from some 10 keV up to some 100 keV, the penetration depth is sufficiently large to probe bulk volumes (some 100 nm). Surface effects become small, and the interpretation of spectra is simplified since the fast probe electron and the (slow) target electrons can be considered distinguishable, hence, exchange effects can be safely ignored.

From the previous, one may infer that inelastic electron scattering, although an important physical phenomenon *per se* is considered primarily a tool to obtain information on matter. It is this— very personal—view which governs the selection and sequence of items in the following sections. Due to the different approaches to inelastic processes (atomic physics and solid state physics) there are different terminologies and conceptions. In processes with energy transfer much higher than the interatomic binding energy, solid state effects may be ignored. Cross sections are then, to a first approximation, dominated by excitation of core levels. They are interpreted in terms such as transition matrix elements or generalized oscillator strength familiar in atomic physics. Chapter 1, 2 and 3 are dedicated to these "core losses". In Chapter 1 we introduce the classical concept of impact parameter, show its applicability to the calculation of cross sections and the eventual breakdown of the classical conception in the Ramsauer-Townsend effect.

Chapter 2 delivers a quantum mechanical foundation of inelastic scattering, mainly following the line of thought elicited by H. Bethe in 1930. We shall discuss the fluctuation-dissipation theorem which relates the structure factor to the density-density correlation function. Chapter 3 accounts for practical aspects of edge spectroscopy.

The different terminology of solid state physics as compared to atomic physics becomes obvious in the interpretation of lower energy transfer. In the range from ~ 1 eV to some 10 eV, energy loss spectra are dominated by transitions from the valence or conduction band. Obviously, such spectra contain information on the density of states up to the continuum and even to higher energies. Unfortunately, there is no direct way of extracting the density of states from spectra. One reason is the occurrence of transition matrix elements in the transition probability which are only known in a few model cases. Even if the influence of matrix elements is ignored, only the joint density of states which corresponds to a convolution of occupied with unoccupied densities of states, can be retrieved from spectra.

Apart from the influence of transition matrix elements, the joint density of states is described by the imaginary part of the dielectric permittivity ε_2, as a function of frequency. Switching from the quantum mechanical joint density of states to the phenomenological description by ε, the difference in energy between particular initial and final states of a transition is viewed as the energy of an oscillator of particular frequency, contributing to the dielectric function ε. It is probably due to the aforementioned impediments in deriving densities of states that low energy losses are still approached in the framework of phenomenological Maxwell theory, similar to optical spectroscopy.

We shall review classical electrodynamics in a form suitable for interpretation of low energy losses in Chapter 4. For a general understanding of the relations between the eigenmodes of a system and the dielectric function, it suffices to assume an isotropic, homogenous and nonmagnetic material. Theory predicts that besides transverse eigenmodes which couple to electromagnetic radiation, the charge carriers in the medium may exhibit longitudinal oscillations if the real part of the dielectric function vanishes at any frequency. These are called collective oscillations or plasmons, in analogy to the notation phonon for a collective lattice oscillation. In metals (which resemble solid state plasmas) collective oscillations are strong and well defined. They show up in electron energy loss spectra. (Not so in optical spectra since longitudinal modes do not couple to transverse radiation.) It was probably their alleged unimportance or undetectability in "classical" spectroscopy which caused nearly complete ignorance of plasmons among solid state physicists for a long time.

Although well-known in classical electrodynamics (early papers - 1920) A. Sommerfeld and H. Bethe do not mention them in their standard monograph "Electron Theory of Metals". Even in university courses on classical electrodynamics, plasmons are likely to be overlooked.

In Chapter 5 some details on charge oscillations are discussed. We investigate the effect of screening, the influence of boundaries on the eigenmodes of a medium, viz. the occurrence of surface oscillations, which make feasible a rather new type of experiment with attenuated total reflection (ATR). From the existence of longitudinal modes an interesting and quite surprising interference as to the validity of Fresnel's Equations will be drawn. Finally, in Chapter 5, we derive an expression for the differential cross section and the energy loss of a probe electron.

It turns out that the dielectric function ε of the medium completely determines the energy loss of the probe electron. Vice versa, ε can be determined by an inelastic scattering experiment.

Chapter 6 is a compendium of basic definitions and theorems of quantum mechanics with emphasis on the many-body theory.

Chapter 6 may be considered an introduction to Chapter 7 which deals with the electron gas as a many-body system. We start with the Sommerfeld non-interacting gas and proceed to better approximations of the interacting gas. A common method to do so is to apply perturbation theory. It turns out, however, that in the quantum mechanical perturbation series divergences arise, no matter in which order the expansion is performed. Not till 1955, when D. Bohm and D. Pines came forth with a canonical transform approach, was the problem solved. Up to now, a number of quantum

mechanical treatments have been set up. The selective summation technique, based upon Feynman diagrams, is one of the most
transparent. The method aims at calculating the system's Green
function which eventually leads to the dielectric function.

The last section includes some details not yet covered. A gen-
eralization of the previously defined dielectric function to periodically inhomogeneous media, and the modification of ε, as already
derived for a free electron gas, by a weak pseudopotential. Chapter
8 also outlines how to retrieve the relevant information from experiments. In particular, a serious masking effect in the spectra is
multiple scattering. Inclusion of coherent multiple scattering is not
yet operational. Existing correction methods rely on the assumption that no phase relations between subsequent inelastic scattering
events exist on the average. On this assumption we present a new
approach to the retrieval of the loss function from experimental
results.

Contents

1. Classical Scattering Theory   1
   1.1   Elastic Cross Sections   1
   1.2   The Deflection Function   4
   1.3   Scattering on a Hard Sphere   5
   1.4   The General Case   6
   1.5   Rutherford Scattering   9
   1.6   Singularities   9
   1.7   Inelastic Cross Sections   10
   1.8   The Ramsauer-Townsend Effect   12
   1.9   On the Validity of the Classical Description 13

2. Quantum Mechanical Scattering Theory   15
   2.1   Absorption Edges   15
   2.2   The Differential Cross Section   16
   2.3   The Dynamic Form Factor   18
   2.4   The Generalized Oscillator Strength . . .   22
   2.5   Rutherford Scattering   23
   2.6   The Bethe Differential Cross Section . .   23
   2.7   The Hydrogenic Approach   26
   2.8   The Bethe Approximation   29
   2.9   Zonal Harmonics Expansion   30
   2.10   Qualitative Interpretation   32
   2.11   The Total Elastic Cross Section   34
   2.12   The Ramsauer-Townsend Effect   38

3. Practical Aspects of Absorption Edge
   Spectroscopy   41
   3.1   A Survey of Applications   41
   3.2   Microanalysis   43
   3.3   Electron Compton Scattering   47
   3.4   Site-specific Excitations   50
   3.5   Extended Energy Loss Fine Structure
          (EXELFS)   52

4. Electrodynamics in Homogeneous,
   Isotropic Media 55
   4.1   Fourier Transform 58
   4.2   Linear Response of a Medium 60
   4.3   The Maxwell Equations 62
   4.4   The Dielectric Function 67
   4.5   The Drude Model 68
   4.6   Charge Oscillations in a Metal 71

5. Some Details on Charge Oscillations . . 75
   5.1   Screening 75
   5.2   Plasmons 78
   5.3   Plasmon Dispersion 80
   5.4   Boundaries 82
   5.5   Surface Oscillations 85
   5.6   The ATR-Method 86
   5.7   On the Restricted Validity
          of the Fresnel Equations 89
   5.8   The Differential Cross Section 92
   5.9   Energy Loss Function 95

6. Quantum Mechanical Preliminaries . . . 98
   6.1   Summary of Important Facts 99
   6.2   The Lippman-Schwinger Equation . . . 101
   6.3   The Green Operator 103
   6.4   The Dyson Equation 105
   6.5   Green Operators in the Time Domain . . 106
   6.6   Relation of G_0 to the Time Evolution
          Operator 108
   6.7   Second Quantization 109
   6.8   Operators in Second Quantization 111
   6.9   The Perturbation Series
          in Graphical Representation 114
   6.10   An Example 116
   6.11   The Coulomb Interaction 119

7. Quantum Mechanical Description
   of the Electron Gas   123
   7.1   The Jellium Hamiltonian   123
   7.2   Sommerfeld Non-interacting e^--Gas . . .   125
   7.3   Hartree Approximation (HA)   127
   7.4   Hartree-Fock Approximation (HFA) . . .   130
   7.5   Why Higher Order Perturbation Theory
         Does not Work   132
   7.6   The Random Phase Approximation (RPA)   135
   7.7   Electron Scattering   141
   7.8   Polarization Diagrams   142

8. Beyond Simple Models   144
   8.1   Solid State Effects   144
   8.2   Ion-Electron Interactions   149
   8.3   Multiple Scattering   152

References   167

Author Index   174

Subject Index   177

1. Classical Scattering Theory

The Rutherford model [1.1] of the atom consisting of a positively charged, virtually pointlike nucleus surrounded by a cloud of negative charges was founded on a classical scattering theory. Alpha particles were considered pointlike, moving under the influence of a Coulomb force. Experimental results were completely explained by the angular distribution of scattered particles obtained from their classical trajectories. In 1911, when the classical model came forth, no one knew about wave mechanics or probability amplitudes. Nor did anyone know that there is nothing like a trajectory on the atomic scale. Though the theory was successful— today we know it was good luck, since the classical and quantum mechanical cross sections coincide, by chance, for the Coulomb potential. So, it can be said that a wrong theory, applied onto a correct model, made an early emergence and an early success of atomic physics possible.

Still today, classical concepts have their part in scattering theory, because of their clarity and because they may be applied in particular cases much easier than quantum mechanics. We shall, therefore, outline the fundamental concepts leading to the prediction of scattering cross sections from a model potential. In doing so, we shall consider elastic interactions first, and extend them to inelastic ones.

1.1. Elastic Cross Sections

Scattering of a probe particle on a target is said to be elastic if there is no change in the *internal* energy of both. The maximum ΔE of kinetic energy the probe particle can transfer to a target at rest is given by the conservation of energy E_0 and momentum $\vec{p}_0$ in a head-on elastic collision:

$$p_0 = p_1 + p_2 \tag{1.1a}$$

$$E_0 = \frac{p_0^2}{2m} = \frac{p_1^2}{2m} + \frac{p_2^2}{2M} \tag{1.1b}$$

$$\Delta E = \frac{p_2^2}{2M} = \frac{4mM}{(M+m)^2}E_0 \doteq 4\frac{m}{M} \cdot E_0 \tag{1.1c}$$

where $m \ll M$ are the masses of the probe (an electron) and the target, resp. (see Fig. 1.1 for collision kinematics).

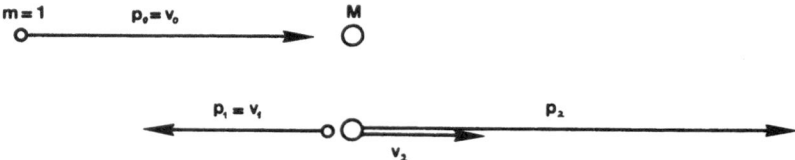

Fig. 1.1. Kinematics of head-on collisions.

Since the target consists of atoms we have for the mass ratio $m/M \leq 10^{-4}$, and

$$\Delta E \leq 10^{-4}E_0. \tag{1.2}$$

For molecules, ΔE is even smaller by a factor of 10 to 1000, and for crystals, where $\sim 10^{23}$ atoms are tightly connected to a lattice, $m/M \approx 10^{-27}$. Hence, except for the lightest atoms in the gas phase, the elastic energy transfer in scattering of electrons on matter is negligibly small, and one can identify elastic interactions as those in which the energy of the probe electron is practically conserved. (An equivalent statement is that the "center of mass" system of the collision coincides with the "laboratory" system for elastic scattering of electrons on matter.)

The elastic differential cross section $\partial^2\sigma/\partial\vec{\Omega}$ is defined as the area which is *effective* per atom for scattering through solid angle $\vec{\Omega}$. The number of scattered particles $\Delta N_S(\vec{\Omega}, \Delta\Omega)$ within a small solid angle $\Delta\Omega$ around $\vec{\Omega}$ which the detector spans is then

$$\Delta N_S(\vec{\Omega}, \Delta\Omega) = I_0 \frac{\partial^2\sigma}{\partial\vec{\Omega}}\Delta\Omega N = I_0\Delta f(\vec{\Omega}) \cdot N \tag{1.3a}$$

where I_0 is the intensity of the incident beam and N the number of (irradiated) atoms in the thin specimen (see Fig. 1.2). Since the intensity of the scattered particles is related to ΔN_s as

$$I_S(\vec{\Omega}) \cdot \Delta \Omega R^2 = \Delta N_S(\vec{\Omega}, \Delta \Omega) \tag{1.3b}$$

we obtain

$$\frac{\partial^2 \sigma}{\partial \vec{\Omega}} = \frac{I_S(\vec{\Omega})}{I_0} \frac{R^2}{N} \tag{1.4}$$

from measurement of I_s, where R^2 and N is known.

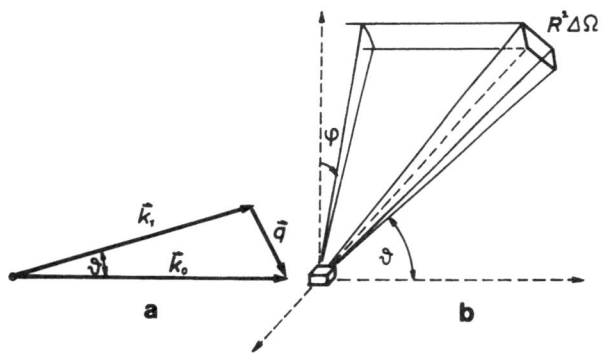

Fig. 1.2. a) Scattering plane with wave vectors of incident beam $(\vec{k}_0)$ and scattered beam $(\vec{k}_1)$. $\vec{q}$ is the wave vector transferred to the target.
b) Experiment for measuring angular differential cross section. The detector area is $R^2 \Delta \Omega$.

The scattering probability is

$$\frac{\partial^2 p}{\partial \vec{\Omega}} = \frac{\partial^2 \sigma}{\partial \vec{\Omega}} \frac{N}{F_0} = \frac{\partial^2 \sigma}{\partial \vec{\Omega}} nd. \tag{1.5}$$

Here the necessity of *thin* targets becomes obvious, since the probability cannot exceed 1. Strictly speaking Eqs. (1.3) - (1.5) are only valid in infinitely dilute targets where single atoms do not "shadow" one another. To put it another way: multiple scattering was neglected in (1.3) - (1.5).

Integrating (1.5) over $\vec{\Omega}$, we have

$$p = \sigma nd. \tag{1.6}$$

In a target of thickness $\lambda = (\sigma n)^{-1}$, which is called the mean free path, each electron has suffered one collision, and shadowing effects become dominant. We conclude that "thin" means $d \ll \lambda$. When $d \geq \lambda$, multiple scattering effects have to be accounted for. It is easy to see what happens to the sum total of scattered electrons, $I = \sum_{\Delta\Omega} \Delta N_s$. In an infinitely thin layer of the target, Eq. (1.6) is $\delta p = \sigma \cdot n \cdot \delta d$ which yields, by scaling to the intensity,

$$\frac{\partial I(d)}{\partial d} = I_0(d)\sigma \cdot n. \tag{1.7}$$

Due to conservation of particles, $I(d) + I_0(d) = I_0(0)$, and

$$I_0(d) = I_0(0)\exp(-\sigma nd). \tag{1.8}$$

The unscattered beam is attenuated exponentially with increasing thickness, the mean free path being the penetration depth.

1.2. The Deflection Function

The aim of scattering theory is to calculate $\partial^2\sigma/\partial\vec{\Omega}$ from model potentials so as to compare them with experiments. A useful concept in doing so is the impact parameter (Fig. 1.3). An electron approaching a target atom at a perpendicular distance b (the impact parameter) is deflected by an angle $\vartheta(b)$. From Fig. 1.3, for cylindrically symmetric scattering,

$$d\sigma = 2\pi bdb = \int_0^{2\pi} \frac{\partial^2\sigma}{\partial\vec{\Omega}} d\varphi \sin\vartheta d\vartheta \tag{1.9}$$

and hence

$$\frac{\partial^2\sigma}{\partial\vec{\Omega}} = \left| \frac{b}{\sin\vartheta} \frac{db}{d\vartheta} \right| = \left| \frac{b(\vartheta)}{\sin\vartheta} \frac{1}{d\vartheta/db} \right|. \tag{1.10}$$

The central problem is to obtain the classical deflection function $\vartheta(b)$ from a given model potential.

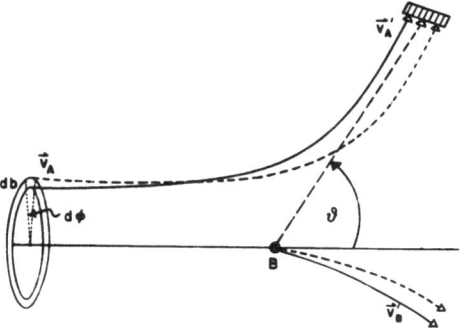

Fig. 1.3. Collision of particles A and B. All particles A passing through the ring have similar trajectories and experience very nearly the same deflection ϑ. From [1.3].

1.3. Scattering on a Hard Sphere

The deflection function for scattering of a pointlike particle on a hard sphere of radius a is

$$b(\vartheta) = a \sin \frac{\pi - \vartheta}{2} = a \cos \frac{\vartheta}{2} \qquad (1.11)$$

as can be seen from Fig. 1.4. Hence, from (1.10),

$$\frac{\partial^2 \sigma}{\partial \vec{\Omega}} = a \frac{\cos \vartheta/2}{\sin \vartheta} \cdot \frac{a \sin \vartheta/2}{2} = \frac{a^2}{4} \qquad (1.12)$$

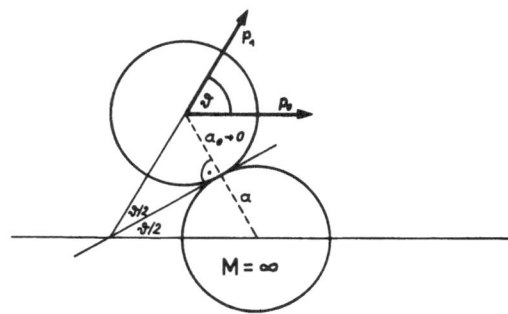

Fig. 1.4. Scattering on a hard sphere of infinite mass.

which is to say: The scattering on a hard sphere is isotropic. The total classical cross section is given, by integration over solid angle 4π, as $\sigma = a^2\pi$ which is the projected area of the sphere.

1.4. The General Case

The general case of scattering off a central potential $V(r)$ is treated by solving the equation of motion of the probe particle in the central field. It is helpful, in doing so, to write the total energy in terms of the radial and angular momenta p_r and L:

$$E = E_{kin} + V(r) = \frac{p_r^2(r)}{2m} + \frac{L^2}{2mr^2} + V(r) \qquad (1.13a)$$

$$L := p_\alpha \cdot r = mr^2\dot{\alpha} \qquad (1.13b)$$

$$p_r := m\dot{r}. \qquad (1.13c)$$

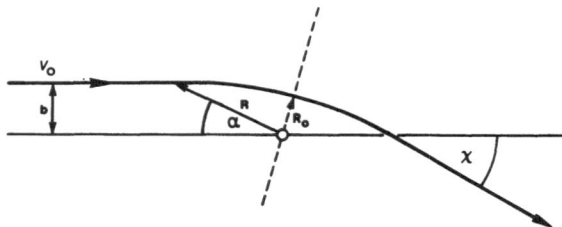

Fig. 1.5. Trajectory for particle with initial velocity v_0 and impact parameter b subject to a central attractive force. The quantity χ is the CM scattering angle, R the distance to the fixed center, α the angular position of the particle along the trajectory, and R_0 the distance of the closest approach. Dashed line indicates symmetry of the collision.

Observe that both E and L are constants of motion. As can be seen from Fig. 1.5,

$$L = mv_0b, \qquad (1.14a)$$

$$E = \frac{mv_0^2}{2}. \qquad (1.14b)$$

It is useful to introduce an "effective" potential

$$V_{eff}(L, r) = \frac{L^2}{2mr^2} + V(r).$$ (1.15)

The total energy Eq. (1.13a) can now be written as the sum over the "radial" kinetic energy and V_{eff}, —so we obtain for the radial momentum:

$$p_r(r) = \sqrt{2m(E - V_{eff}(L, r))}.$$ (1.16a)

Important characteristics of the particle's motion can be visualized graphically (see Fig. 1.6, which is a diagram of $V_{eff}(r)$ for three different impact parameters. The potential energy V is a Lennard-Jones-six, twelve-potential as it occurs in neutral symmetric systems, e.g. in rare gases.). The distance of closest approach is given by the demand that $E - V_{eff} \geq 0$ in order that p be real. Obviously $L^2/2mr^2$ acts as a centrifugal barrier which the particle cannot penetrate. For particular values of L, closed orbits are possible. In general, they cannot be approached from infinity, i.e. in scattering experiments, particle trapping does not occur. In the case that the local maximum of V_{eff} coincides with E, a particle approaching from infinity may reach the critical distance r_c. Eq. (1.16) tells that the radial velocity vanishes at r_c, and so does the radial acceleration

$$\dot{v}_r = \frac{1}{m}\dot{p}_r = \frac{1}{m}\frac{dp_r}{dr}\frac{p_r}{m}.$$ (1.16b)

Physically, this means that the particle orbits the scatterer in a circular trajectory. The deflection function $\vartheta(b)$ will tend to infinity at a critical impact parameter. The orbit is instable, since a small perturbation leads to either trapping or receding. —Orbiting can occur only when the attractive part of $V(r)$ falls faster than $1/r^2$.

In passing we mention that the same line of thought is applied in astronomy, concerning trajectories in a central field. General relativity yields an attractive potential $\propto 1/r^3$ superimposed onto the Newtonian gravitation, hence orbiting may occur. The critical distance is 1.5 Schwarzschild radii, far inside the gravitating bodies

8

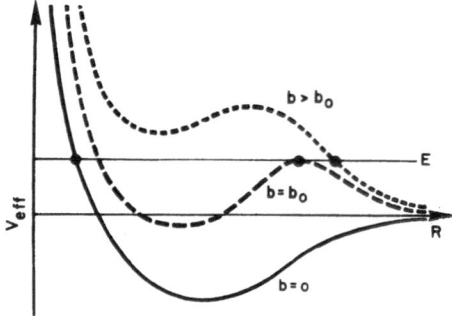

Fig. 1.6. Effective potential, $V_{eff} = V(R) + L^2/2mR^2 = V(R) + Eb^2/R^2$ as a function of distance R, for three impact parameters (or angular momenta), with CM energy $E : b = 0$, $V_{eff} = V(R)$; $b = b_0$, orbiting occurs at distance of closest approach; $b > b_0$. Distances of closest approach indicated by dots. From [1.3].

except for black holes. A spacecraft, targeted at a black hole with the proper impact parameter may stay in an orbit forever [1.2].

The total deflection angle of the probe particle is, according to Fig. 1.5,

$$\vartheta = \pi - \int\limits_{-\infty}^{+\infty} \dot\alpha \, dt = \pi - \int\limits_{-\infty}^{+\infty} \frac{v_0 b}{r^2(t)} dt. \qquad (1.17)$$

Now, substitute $dt = dr/\dot r$ and split the integral into 2 equal parts, so as to obtain from Eqs. (1.16), (1.17)

$$\vartheta = \pi - 2b \int\limits_{r_{min}}^{\infty} \frac{dr}{r^2 \sqrt{1 - \dfrac{V_{eff}(L,r)}{E}}}. \qquad (1.18)$$

For large impact parameters, when $V \ll E$ we have $V_{eff}/E \doteq$ $\doteq L^2/(2mr^2E)$. In this case, called classical impulse approximation, the deflection can be evaluated analytically. In the general case, however, Eq. (1.18) is, for most potentials, evaluated numerically.

1.5. Rutherford Scattering

In case of the repulsive Coulomb potential between, say, two electrons of charge e, the integral is analytic, yielding

$$\vartheta = 2arc\sin(1/\sqrt{1 + mv_0^2 b/e^2}).\qquad(1.19)$$

Inserting Eq. (1.19) into (1.10), the classical Rutherford cross section is obtained:

$$\frac{\partial^2 \sigma}{\partial \vec{\Omega}} = \left(\frac{e^2}{2mv_0^2}\right)^2 \frac{1}{\sin^4 \vartheta/2}.\qquad(1.20)$$

Due to the infinite range of the Coulomb field, a singularity at $\vartheta = 0$ is apparent. Real fields are always of finite extent because of screening (see Chapter 5), so the singularity is removed.

Incidentally, the classical Rutherford cross section for Coulomb scattering coincides with the quantum mechanical formula [1.3], [1.4], [1.5]. This incidence was the very reason for the early successes and the rapid development of atomic physics.

1.6. Singularities

In general, the deflection may be a complicated function of b (see Fig. 1.7 for a Lennard-Jones potential $V(r) \sim (a/r^{2n} - b/r^n)$). The insert on the left of the figure shows an extremum in deflection at b_r, so $d\vartheta/\partial b = 0$. Hence, from (1.10), $\partial^2 \sigma/\partial \vec{\Omega} \to \infty$. The cross section is singular at ϑ_r. Since this effect also causes rainbows by reflection of light in water droplets ($\vartheta_r \approx 138°$ depending on colour) a singularity of this type is called rainbow singularity.

Inspection of Eq. (1.10) shows that for $\vartheta = 0$ or $\vartheta = \pi$ singularities may occur, too. They are known as forward and backward glory singularities because they are responsible for the meteorological phenomena of glory effects in a foggy environment. The backward glory effect is also exploited by embedding small spherical particles into a transparent medium, for the production of highly reflecting stickers or traffic signs.

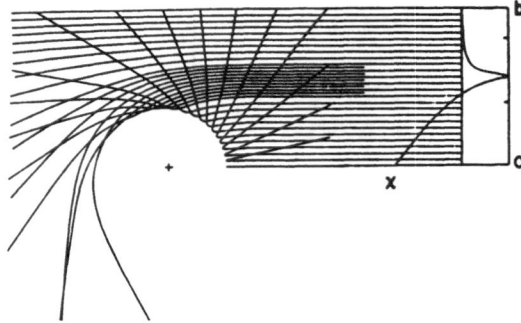

Fig. 1.7. Classical trajectories corresponding to a collision of a particle with a fixed force center. Interaction is a long-range attractive and short-range repulsive potential. Deflection function χ vs b is shown on the right. From [1.10].

1.7. Inelastic Cross Sections

In an inelastic interaction, the *internal* energy of the specimen is changed, the change being equal to the energy loss (or gain) of the probe particle. Since in inelastic collisions not only the direction of flight of the probe particle is changed, but also its energy, it is reasonable to define a doubly differential cross section $\partial^3\sigma/\partial E\partial\vec{\Omega}$. Inelastic processes are numerous, e.g. ionisation of atoms, interband transitions of valence or conduction electrons, excitation of phonons or plasmons, etc. Most of these are not included in the classical kinematic theory developed above. Only *collisions* between particles are covered, which is one of the serious deficiencies of the classical theory.

"Inelastic" in the context of classical theory can only mean that some of the constituent particles of the target acquire (or lose) energy in a collision. In the most simple approach, only binary encounters are considered, say, of the probe electron with one of the target electrons assumed to be at rest. Conservation of energy and of momentum yields three scalar equations for four unknown scalar quantities, namely the velocities of the scatterer and the scattered particle in the plane of scattering.

One can eliminate the motion of the scatterer from these equations; there remains one relation between the velocity of the probe particle and the deflection angle ϑ, or, what is equivalent, a relation between ϑ and the energy loss E. Explicitly, [1.3]

$$E = E_t + \frac{4mM}{(M+m)^2}(E_0 - E_t)\sin^2\chi/2 := g(\chi/2) \qquad (1.21)$$

where E_0 is the primary energy of the probe, and E_t is an energy threshold necessary to free the bound scatterer (ionisation energy). m, M are the respective masses of the probe and the scatterer where χ is the deflection angle in the center of mass-system. For small χ, and equal masses of the colliding particles, the transformation to the deflection ϑ in the laboratory system is $\vartheta = \chi/2$, as can be visualized graphically. In this latter form, Eq. (1.21) is often encountered in the literature on Compton scattering [1.6] (cf. Sect. 3.3.).

From (1.21) and (1.10), we find

$$\frac{\partial^3\sigma}{\partial E\partial\vec{\Omega}} = \left|\frac{b(\vartheta)}{\sin\vartheta}\frac{1}{d\vartheta/db}\right|\cdot\delta(E - g(\vartheta)) = \frac{\partial^2\sigma}{\partial\vec{\Omega}}\delta(E - g(\vartheta)). \qquad (1.22)$$

It is straightforward to derive the "energy transfer" cross section

$$\frac{\partial\sigma}{\partial E} = \int\frac{\partial^3\sigma}{\partial E\partial\vec{\Omega}}d\vec{\Omega} = \frac{4\pi}{4mME_0/(M+m)^2}\frac{\partial^2\sigma}{\partial\vec{\Omega}}\bigg|_{g^{-1}(E)}. \qquad (1.23)$$

Contrary to our simplification, the target particles (electrons) will not be at rest, in reality. As a consequence, (1.21) will change, and the differential cross section (1.22), subject to measurements, will be determined from an average over the initial velocity distribution of scatterers. This will result in a smearing of the δ-function in Eq. (1.22) over the (E, ϑ)-plane. In fast collisions, when the primary energy E_0 of the probe is much higher than the initial kinetic energy of the target particles, the broadening effect will be small (see Fig. 1.8). The loci of maximal scattering probabilities are called Compton line.

From the previous considerations, it is apparent that the total cross section decreases with increasing primary energy E_0, which is to say that matter is more transparent for high energy particles, a fact that can be conceived intuitively.

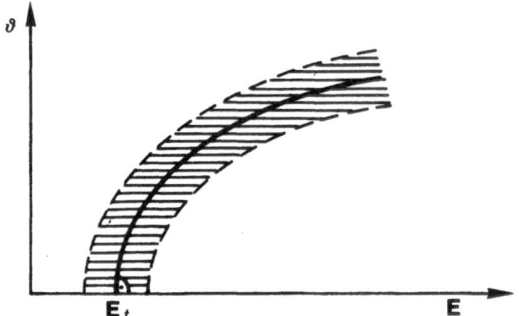

Fig. 1.8. Scattering cross section for moving targets: $\partial^2\sigma/\partial E\partial\Omega \neq 0$ within the shaded region. Along the full line $\partial^2\sigma/\partial E\partial\Omega \neq 0$ for a target at rest.

1.8. The Ramsauer-Townsend Effect

The flaw came in 1921, when the classical model described above was well established. Ramsauer and Townsend independently discovered that the scattering cross section of the rare gases Ar, Kr, Xe for slow electrons decreased with decreasing energy [1.7], [1.8], [1.9]. For very slow electrons (~ 0.7 eV) the gases are nearly transparent, a behaviour which cannot be explained within the classical theory (Fig. 1.9).

Quantum mechanics lateron overcame the conceptual hurdles. The wave properties of the electron are responsible for the Ramsauer-Townsend effect in that the particle wave experiences a phase shift during scattering. At a particular electron velocity (corresponding to 0.7 eV kinetic energy for Ar), the phase shift of the electron wave acts so as to diminish all scattered contributions by destructive interference; hence, the beam of electrons can transmit the gas nearly unchanged. (See Sect. 2.12 for a more detailed discussion).

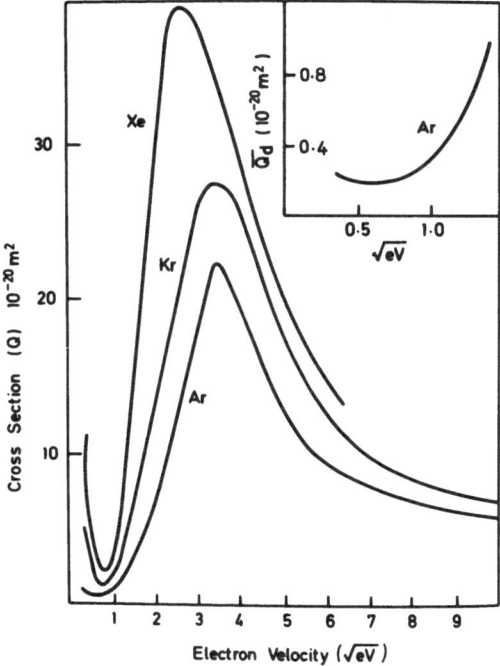

Fig. 1.9. Variation with electron velocity of the total electron collision cross-section in Ar, Kr, and Xe as measured in experiments of the Ramsauer Type. The inset gives the variation of mean diffusion cross-section with root mean square electron velocity in argon, derived from drift experiments . Note that 1 eV corresponds to an electron velocity of $0.6 \times 10^{5} ms^{-1}$. From [1.4].

1.9. On the Validity of the Classical Description

The question arises as to when the classical description for scattering is valid. Intuitively, one can assume that the classical concept of the particle trajectory should be a good approximation in this case. To put it more strictly, the uncertainty Δb in impact parameter should be much smaller than b

$$\Delta b \ll b \leq a \qquad (1.24)$$

where a is the range of the scattering potential; and the scattering angle ϑ should be well defined:

$$\Delta \vartheta \ll \vartheta \qquad (1.25)$$

which is equivalent to (see Fig. 1.3)

$$\Delta v_y \ll v\vartheta \qquad (1.26)$$

The uncertainty principle for the y-component of motion reads

$$\Delta b \cdot \Delta v_y \cdot m \simeq \hbar. \qquad (1.27)$$

Using inequalities (1.24), (1.26) in (1.27) yields

$$\hbar \ll av\vartheta m = a\vartheta\hbar k \qquad (1.28)$$

or

$$\vartheta \gg \lambda/2\pi a. \qquad (1.29)$$

Which is to say: the smaller the wavelength of the quantum mechanical wave train (the higher the particle's velocity), measured on the scale a of the scattering potential, the smaller the angle at which quantum mechanics is necessary to describe scattering. It will be shown in the next section that the quantum mechanical scattering cross section departs from the classical one, the differences mainly occurring within angles smaller than given by Eq. (1.29).

2.Quantum Mechanical Scattering Theory

2.1. Absorption Edges

In this chapter we are concerned with excitations from inner shells of atoms. They give rise to absorption edges in Energy Loss Spectroscopy (ELS) similar to the well-known X-ray absorption edges (Fig. 2.1). In general these edges occur at energies $E \geq 50$ eV for most elements, which is far beyond the interaction energy of valence electrons in the solid state. So, it is justified to treat these target electrons as bound to single nuclei, not interacting with other atoms. (The fact that the wave functions of inner shell electrons of adjacent atoms show negligible overlap is indicative for their relative isolation. However, even a K-shell electron in Na, strongly bound to the nucleus, has a mean dwell time of 5 days at a particular site [2.2]—so there is, in fact, faint interaction).

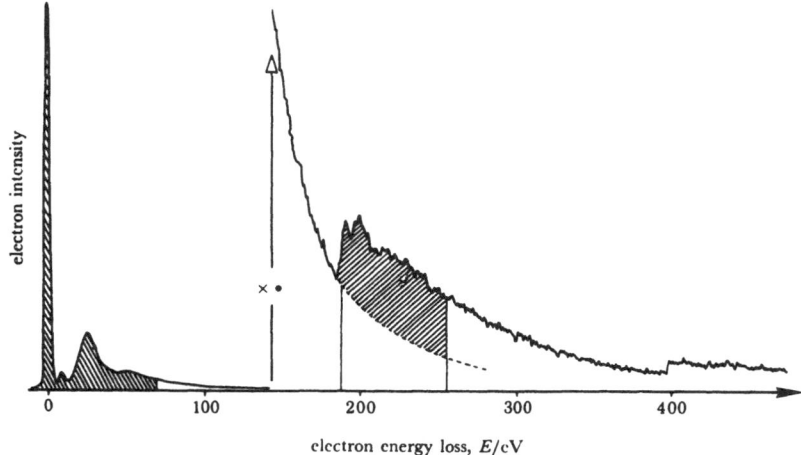

Fig. 2.1. Energy-loss spectrum of boron nitride showing the zero-loss peak, plasmon peak, a rapid gain increment (•-436) and K-shell ionization edges of boron and nitrogen. The hatched areas represent integrated intensities necessary for microanalysis. From [2.19].

Valence and/or conduction electrons, on the other hand, interact rather strongly with one another. Their influence shows up mainly in the low energy loss region ($E \leq 20eV$), which will be discussed in Chapters 4, 5 and 7. In the following, we aim at calculating the probability for inelastic scattering of electrons on atoms, as a function of energy and momentum transfer. Since the latter is related to the scattering angle ϑ (see Fig. 1.2) the scattering probability is the differential cross section $\partial^2\sigma/\partial E\partial\Omega$ which is measurable directly in energy loss experiments. This probability obviously depends on the nature of the scatterer, hence measuring $\partial^2\sigma/\partial E\partial\Omega$ yields information on the target atoms. (Since we do not explicitly assume that the cross section depends on the azimuthal angle, we shall write $\partial^2\sigma/\partial E\partial\Omega$ instead of $\partial^3\sigma/\partial E\partial\vec{\Omega}$ throughout this chapter).

2.2. The Differential Cross Section

A convenient approach starts with the golden rule of first order perturbation theory: (e.g. [2.3])

$$dW_{ba} = \frac{2\pi}{\hbar}| < b|V|a > |^2 dv_b \cdot \delta(E_a - E_b) \qquad (2.1)$$

where

dW_{ba} number of transitions from unperturbed state $|a>$ to state $|b>$ within dv_b around $|b>$ per unit time (transition rate)

V   perturbing potential

dv_b differential phase space element around $|b>$ and

$|b>, |a>$ are unperturbed (usually considered free) states.

Now, the differential particle current of free electrons scattered into $(d\Omega, dE)$ is given by (2.1), summing over final states $|b>$

$$dj = \frac{\partial W}{\partial\Omega\partial E}d\Omega dE = dj(dE, d\Omega) =$$
$$= \sum_b \frac{2\pi}{\hbar}| < b|V|a > |^2\frac{k_b m}{\hbar^2}d\Omega dE \cdot \delta(E_a - E_b) \qquad (2.2)$$

since $dv_b = k_b^2 dk_b d\Omega$ and $E = \hbar^2 k^2 / 2m$ for free states $|b>$.

On the other hand,

$$dj = d\sigma(E, \Omega) \cdot j_i \qquad (2.3)$$

where $d\sigma = dEd\Omega \cdot \partial^2\sigma / \partial E \partial \Omega$ is the scattering cross section and j_i the incident particle current density. From (2.2) and (2.3):

$$\frac{\partial^2 \sigma}{\partial E \partial \Omega} = \sum_b \frac{2\pi k_b m}{\hbar^3 j_i} | < b|V|a > |^2 \delta(E_a - E_b). \qquad (2.4)$$

The incident electrons are free particle eigenstates $|k>$ normalized such that $< k'|k >= \delta^3(\vec{k} - \vec{k}')$

$$< \vec{r}|\vec{k} >= (2\pi)^{-\frac{3}{2}} e^{i\vec{k}\vec{r}} = \varphi(\vec{r}). \qquad (2.5)$$

The quantum mechanical current density is [2.3]

$$j_i = \frac{\hbar}{2mi}(\varphi^* \nabla \varphi - \varphi \nabla \varphi^*), \qquad (2.6)$$

φ from Eq. (2.5), hence $|\varphi >= |k_a >$,

$$j_i = \frac{\hbar k_a}{(2\pi)^3 m} \qquad (2.7)$$

and

$$\frac{\partial^2 \sigma}{\partial E \partial \Omega} = \left(\frac{2\pi}{\hbar}\right)^4 m^2 \sum_b \frac{k_b}{k_a} | < b|V|a > |^2 \delta(E_a - E_b). \qquad (2.8)$$

In elastic scattering, $|b>$ and $|a>$ are considered free states as mentioned, and V is a static potential. This implies energy conservation of the probe.

Inelastic scattering can be treated similarly when V represents the interaction potential between the probe and the target charges. Then, $|a>$ and $|b>$ in Eq. (2.8) resemble many-particle state vectors, including the target electrons.

2.3. The Dynamic Form Factor

Eq. (2.1) is valid when V is small compared with the energy of the scattered electrons, i.e. for fast collisions [2.11]. In this case, the unperturbed state vectors can be approximately factorized since exchange effects are negligible (so to speak, the fast electron and the atom electrons are distinguishable):

$$|a> = |k_a > \otimes |\varphi > \qquad |b> = |k_b > \otimes |\varphi' >, \qquad (2.9)$$

k_a, k_b designating free states, φ, φ' are the eigenfunctions of the electrons bound to the target nucleus,—before and after the collision. In general, φ is, in $\vec{r}$-space: $\varphi(\vec{r}_1, \ldots \vec{r}_N)$ for an N-electron atom. Hence, $< b|V|a >$ is obtained in $\vec{r}$-representation by inserting

$$1 = \int d^3 r_1 \ldots d^3 r_N |r > \otimes |r_1 \ldots r_N >< r_1 \ldots r_N |\otimes < r|d^3 r. \qquad (2.10)$$

We mention that this definition of the unity operator demands the normalization (2.5).

By virtue of (2.9) and (2.5)

$$< b|V|a > = \int d^3 r_1 \ldots d^3 r_N \varphi'^*(\vec{r}_1 \ldots \vec{r}_N) \varphi(\vec{r}_1 \ldots \vec{r}_N)$$

$$d^3 r V(\vec{r}; \vec{r}_1 \ldots \vec{r}_N) \frac{e^{i(\vec{k}_a - \vec{k}_b)\vec{r}}}{(2\pi)^3}. \qquad (2.11)$$

V is the interaction potential between probe electron at position $\vec{r}$ and one atom (since the target is an arrangement of non-interacting atoms, as long as inner shell excitations are considered):

$$V = \frac{-Ne^2}{r} + \sum_1^N \frac{e^2}{|\vec{r} - \vec{r}_i|}. \qquad (2.12)$$

The $d^3 r$-integration in Eq. (2.11) can be performed by substituting $\vec{r}' = \vec{r} - \vec{r}_i$ in (2.12) and using the Fourier transform of the Coulomb potential

$$\int d^3 r \frac{e^2}{|\vec{r} - \vec{r}_i|} e^{i\vec{q}\vec{r}} = \frac{4\pi e^2}{q^2} e^{i\vec{q}\vec{r}_i}, \qquad (2.13)$$

where $\vec{q} = \vec{k}_a - \vec{k}_b$ is the wave vector transferred during interaction. Eventually, one obtains from Eqs. (2.8), (2.11), (2.12), (2.13):

$$\frac{\partial^2 \sigma}{\partial E \partial \Omega} = \left[\frac{2me^2}{(\hbar q)^2} \right]^2 \frac{k_b}{k_a} |S|^2$$

$$|S|^2 := \sum_{\varphi'} | < \varphi'| \sum_i^N e^{i\vec{q}\vec{r}_i} |\varphi> - N \underbrace{< \varphi'|\varphi >}_{0} |^2 \cdot$$

$$\cdot \delta(E_\varphi - E_{\varphi'} + E) =$$

$$= \sum_{\varphi'} [\int d^3 r_1 \ldots d^3 r_N \varphi'^*(r_1 \ldots r_N) \sum_i^N e^{i\vec{q}\vec{r}_i} \varphi(r_1 \ldots r_N)]^2 \cdot$$

$$\cdot \delta(E_\varphi - E_{\varphi'} + E).$$

(2.14)

where the energy loss $E = E_{k_a} - E_{k_b}$ has been introduced. The last term in Eq. (2.14) which is due to the Coulomb force of the nucleus, vanishes since $|\varphi'>, |\varphi>$ are orthogonal. S is termed inelastic or dynamic scattering form factor or structure factor well known in nuclear physics and solid state physics. Some authors use the notation S for $|S|^2$.

Note that the scattering cross section for small wave number transfers is enhanced by appearance of a factor q^{-4}. One can give the dynamical structure factor $|S|^2$ a physical foundation by observing that $\sum_j e^{-i\vec{q}\vec{r}_j}$ is the Fourier transform of the electron density operator $n(\vec{r})$ of the target:

$$n_q := \int d^3 r e^{-i\vec{q}\vec{r}} n(r) = \int d^3 r e^{-i\vec{q}\vec{r}} \sum_j \delta(\vec{r} - \vec{r}_j) = \sum_j e^{-i\vec{q}\vec{r}_j}.$$

(2.15)

Use of (2.15) and replacement of the Delta-function

$$\delta(E_a - E_b + E) = \frac{1}{2\pi\hbar} \int dt e^{i(E_\varphi - E_{\varphi'} + E)t/\hbar}$$

(2.16)

in (2.14) yields

$$
\begin{aligned}
|S|^2 &= \frac{1}{2\pi\hbar} \sum_{\varphi'} \int dt\, e^{iE_\varphi t/\hbar} < \varphi|n_{\vec{q}}|\varphi' > e^{-iE_{\varphi'} t/\hbar} \\
& \quad < \varphi'|n_{-\vec{q}}|\varphi > e^{iEt/\hbar} = \\
&= \frac{1}{2\pi\hbar} \int dt\, e^{iEt/\hbar} < \varphi|e^{iHt} n_{\vec{q}} e^{-iHt} n_{-\vec{q}}|\varphi >= \\
&= \frac{1}{2\pi\hbar} \int dt\, e^{iEt/\hbar} < n_{\vec{q}}(t) n_{-\vec{q}}(0) >,
\end{aligned}
\tag{2.17}
$$

where we have used the facts that a) $n_{\vec{p}}^+ = n_{-\vec{q}}$ since $n(\vec{r})$ is real, b) the set of final states $|\varphi' >$ is complete, hence $\sum_{\varphi'} |\varphi' >< \varphi'| = 1$, and c) the time dependent Heisenberg operator $n_{\vec{q}}(t)$ is defined as $e^{iHt} n_{\vec{q}} e^{-iHt}$, and H is the Hamiltonian of the target [2.4].

We can rewrite $n_{\vec{q}} n_{-\vec{q}}$ as a double Fourier transform with respect to $\vec{r}, \vec{r'}$ (the t-variables and vector symbols are omitted for carity)

$$
\begin{aligned}
n_q n_{-q} &= \int d^3 r' \int d^3 r\, e^{-iq(r-r')} n(r) n(r') = \\
&= \int d^3 r' \int d^3 x\, e^{-iqx} n(x+r') n(r') = \\
&= \int d^3 x\, e^{-iqx} \int d^3 r'\, n(x+r') n(r').
\end{aligned}
\tag{2.18}
$$

The rightermost integral is the density autocorrelation function. Insertion of (2.18) into (2.17) yields the ever-cited theorem that the dynamical structure factor $|S(q,\omega)|^2$ is the space-time Fourier transform of the density autocorrelation.

A relation between correlation functions and structure factor was surmised back in 1910 by Einstein in quite another context, namely the explanation of opalescence in fluids as a scattering effect of electromagnetic waves caused by stochastic density fluctuations in the medium [2.18]; the complete derivation of (2.17), (2.18) was first given in 1954 by van Hove [2.5].

As will be shown in Sect. 5.9, there is an intimate connection between the differential cross section (2.14) and the dielectric permittivity ε of a medium. Hence, there is a relation between ε and the structure factor S. Since the former describes *dissipation* of energy and the latter expresses charge-density *fluctuations* in space and time, this relation is known as fluctuation-dissipation theorem. A complete derivation can be found in [2.4].

It is obvious from (2.17) and (2.18) that inelastic scattering of electrons measures the temporal and spatial variations of the charge density correlation. As such, it contains information on the charge density variations themselves. However, the charge density cannot be retrieved unambiguously from measurements since different density profiles may deliver the same correlation function. The lost information is hidden in the phase of S which cancels when measuring $|S|^2 = SS^*$. Here we encounter the phase problem which is well known in diffraction physics (see for instance [2.6]).

Before proceeding we remark that inelastic scattering of photons is described by a similar formula as (2.14)

$$\frac{\partial^2 \sigma}{\partial E \partial \Omega} = \left[\frac{e^2}{m^* c^2} \right]^2 \frac{E_b}{E_a} \cdot (\vec{e}_0 \cdot \vec{e}_1) |S(\vec{q}, \omega)|^2. \qquad (2.19)$$

m^* is an effective mass of the target electrons, $\vec{e}_0, \vec{e}_1$ are polarisation vectors of the light waves [2.4]. Note that, contrary to (2.14), small wave vector transfers don't appear enhanced by a factor q^{-4}, as was the case in electron scattering. Due to the prefactor—the Rayleigh-Thompson scattering cross section–expression (2.19) is about five orders of magnitude smaller than the cross section for electrons (2.14). Accordingly, count rates are extremely low in photon scattering experiments (e.g. [2.20]). See also Chapter 3.

It is worth mentioning that the general result (2.18) is, first of all, used in the investigation of many-body-effects, such as collective excitations in plasmas. This is due to the fact that Eq. (2.18) does not explicitly relate to wave functions (which are difficult to obtain in many-body systems) whereas (2.14) does. The calculation of absorption edges where many-body-effects do not play a dominant role, is most often directly based on (2.14).

2.4. The Generalized Oscillator Strength

As in optical absorption spectroscopy, it is also common to introduce an "oscillator strength" for the description of absorption edges in electron scattering. Whereas in optical spectroscopy the wave number transferred from the absorbed photon to the specimen is negligibly small, it is *not* in inelastic electron scattering. Hence, the optical dipole oscillator strength which does not depend on q,

$$f_{\varphi\varphi'} = \frac{2m(E_\varphi - E_{\varphi'})}{\hbar^2}| < \varphi'|r|\varphi > |^2 =$$
$$= \frac{2m(E_\varphi - E_{\varphi'})}{e^2\hbar^2}| < \varphi'|er|\varphi > |^2 \qquad (2.20)$$

has to be generalized. A natural choice is the definition

$$f(\vec{q}, E) = \frac{2mE}{\hbar^2 q^2}|S(\vec{q}, E)|^2, \qquad (2.21)$$

termed generalized oscillator strength (GOS). E is the difference in energy of the final and initial state. By expansion of the exponential in Eq. (2.14) one can readily show that in the limit of small q

$$\lim_{q \to 0} f(q, E_{\varphi\varphi'}) = \sum_{\varphi'} f_{\varphi\varphi'} \delta(E + E_\varphi - E_{\varphi'}). \qquad (2.22a)$$

Since the dipole oscillator strength $f_{\varphi\varphi'}$ remains finite, it follows from (2.21) that

$$\lim_{q \to 0} |S(\vec{q}, E)|^2 \propto q^2 = 0. \qquad (2.22b)$$

Now we can compare the small q-behaviour of scattering cross sections for electrons (2.14) and photons (2.19). With decreasing q, the former *increases* as q^{-2}, whereas the latter *decreases* as q^2.

2.5. Rutherford Scattering

In the limit of small energy losses, $\Delta E \ll E$ as compared to the primary energy E of the electron beam, we have, for $q > 2q_{min} = 2E/k_0$

$$q \approx 2k_a \sin \vartheta/2.$$

(see Fig. 1.2b) Inserted in Eq. (2.14), this yields

$$\frac{\partial^2 \sigma}{\partial E \partial \Omega} \approx \frac{\left(\frac{me^2}{\hbar^2}\right)^2}{4k_a^4 \sin^4 \frac{\vartheta}{2}} |S|^2 = \frac{1}{4a_0^2 k_a^4} \frac{1}{\sin^4 \frac{\vartheta}{2}} \cdot |S|^2. \qquad (2.23)$$

where $a_0 = 0.053 \ nm$ in the prefactor is the Bohr radius. The first term is recognized as the classical Rutherford cross section Eq. (1.20) which is also obtained quantum mechanically for scattering by a pointlike charged particle (Coulomb scattering). Hence, the atomic form factor S can be viewed as causing the cross section to depart from the "pointlike" static case.

Eq. (2.23) clearly shows that the scattering interaction can be factorized into a probe dependent (k_a) and an atom dependent part (S). This allows for straightforward interpretation of inelastic scattering experiments, as long as the assumption of distinguishable particles holds, i.e. for fast probes [2.7].

2.6. The Bethe Differential Cross Section

In practically all cases $\partial^2 \sigma/\partial E \partial \Omega$ is symmetric with respect to the direction of incidence. So, the integration over the azimuthal angle can be carried out. It is convenient to use the variable $\ln(qa_0)^2 = \ln Q$ instead of the remaining angular variable ϑ in the cross section (2.14). By virtue of relations

$$d\Omega = \sin \vartheta d\vartheta \int_0^{2\pi} d\varphi, \qquad (2.24)$$

$$q^2 = k_a^2 + k_b^2 - 2k_a k_b \cos \vartheta \qquad (2.25)$$

(see Fig. 1.2a) one obtains

$$\frac{\partial \Omega}{\partial q} = \frac{2\pi q \sin \vartheta}{k_a k_b \sin \vartheta}. \qquad (2.26)$$

For brevity we write $Q := (qa_0)^2$, and

$$\frac{\partial \Omega}{\partial \ln Q} = \frac{\pi Q}{k_a k_b a_0^2}. \qquad (2.27)$$

Eqs. (2.14), (2.27) yield

$$\frac{\partial^2 \sigma}{\partial E \partial \ln Q} = \frac{4\pi}{k_a^2 Q} |S|^2. \qquad (2.28)$$

This can be rewritten, using Eq. (2.21), as

$$\frac{\partial^2 \sigma}{\partial E \partial \ln Q} = \frac{4\pi}{k_a^2} \cdot \frac{\hbar^2 f(q, E)}{2ma_0^2 E} = e^4 \pi \frac{f(q, E)}{E_0 E} = \frac{4\pi a_0^2 f(q, E)}{(E_0/R) \cdot (E/R)}, \qquad (2.29)$$

where $R = me^4/2\hbar^2 = 13.6 eV$. In this form, $\partial^2 \sigma/\partial E \partial \Omega$ is termed "Bethe differential cross section". The advantage of using Eqs. (2.28) or (2.29) instead of (2.14) is the logarithmic representation. Moreover, the total cross section is simply given as the area under GOS (apart from a factor). From Eqs. (2.28), (2.29)

$$\frac{\partial \sigma}{\partial E} = \frac{4\pi}{k_a^2} \int_{Q_{min}}^{Q_{max}} \frac{|S|^2}{Q} d(\ln Q) = \frac{4\pi a_0^2}{(E_0/R)(E/R)} \int_{Q_{min}}^{Q_{max}} f d(\ln Q). \qquad (2.30)$$

The limits of integration are determined by the kinematics of scattering (Fig. 1.2)

$$Q_{min} = (k_a - k_b)^2 a_0^2,$$
$$Q_{max} = (k_a + k_b)^2 a_0^2. \qquad (2.31)$$

Whereas in classical mechanics there is a unique relation between energy and momentum transfer (cf. Eq. (1.21)) there is none in quantum mechanics. To state it another way, the differential cross section $\partial^2\sigma/\partial E\partial \ln Q$ is, in proportion to the Delta-function $\delta(Q - E(Q))$ classically. Sometimes it is argued that this is due to the simultaneous conservation of energy and momentum during interaction. Actually, the reason is that the classical scatterer is considered to be in a definite state of motion initially. It is easy to see, then, that the simultaneous existence of energy-momentum relations both for the scattered electron and for the target forces a Delta-like behaviour of the scattering cross section (dash-dotted line in Fig. 2.2). In a quantum mechanical system, however, such as an atom, the target is *not* in a definite state of motion (momentum eigenstate). Its initial momentum is a continuous probability distribution uniquely related to its wave function. Consequently the scattering cross section is likewise a continuous distribution, which in a sense, mirrors the initial momentum distribution of the target electrons. Hence, it can be said that quantum effects act to smear the Delta-like profile to the continuous Bethe-surface (Fig. 2.3).

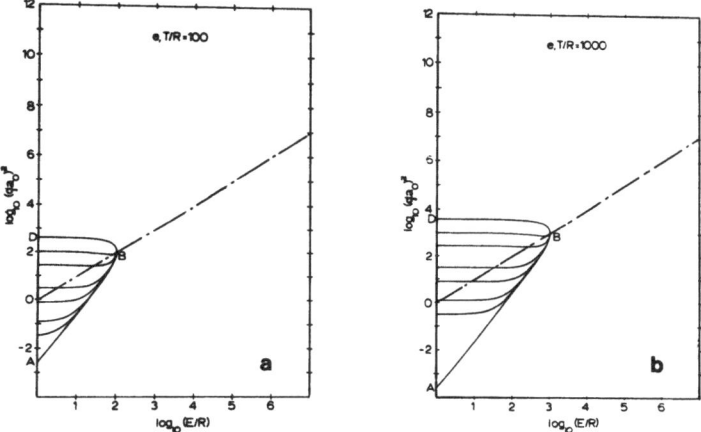

Fig. 2.2. Kinematics for electron impact. Plot (a) refers to $E_0/R = 100$ (1.36 keV kinetic energy) and plot (b) to $E_0/R = 1000$ (13.6 keV kinetic energy). The generalized oscillator strength has substantial magnitude only in the optical region (near the lower left corner) and in the Bethe-ridge region (dash-dotted line). Lines of constant scattering angle are drawn, too. From bottom to top: $\vartheta = 0^o, 1^o, 2^o, 5^o, 10^o, 30^o, 60^o, 180^o$. From [2.7].

What remains valid are the classical upper and lower bounds

of momentum transfer (2.31) at given energy loss. (See Fig. 2.2). The Bethe ridge maximum coincides with the classical δ-like cross section. Also drawn in Fig. 2.2 are lines of constant scattering angle. —For a comprehensive review refer to [2.7], [2.8], [2.9].

2.7. The Hydrogenic Approach

A great deal of work has been done on the calculation of the Bethe cross section for atomic hydrogen, since it is the only system for which GOS is rigorously known. This is due to the fact that the atomic scattering factor S depends on the wave functions of the atom, which are accurately calculable only for hydrogen.

For atomic hydrogen, the GOS happens to be an analytical expression, the calculation of which is straightforward but tedious. We shall not present the formula but rather show some results diagrammatically. (Figs. 2.3, 2.4). Fig. 2.3 has come to be known as "Bethe surface". Due to the relative simplicity of the calculations, attempts have been made to extend the results to the excitation and ionisation of other elements. In this approach, the initial and final wave functions of the atom are assumed to be solutions of the Schrödinger Eq. for hydrogen, scaled to take into account the effective potential of the nucleus and the screening from the outer shells. This widely used concept is called hydrogenic model. Its results are most persuasive for K-shell excitations where hydrogenic wave functions are a good approximation [2.10]. See Fig. 2.5 .

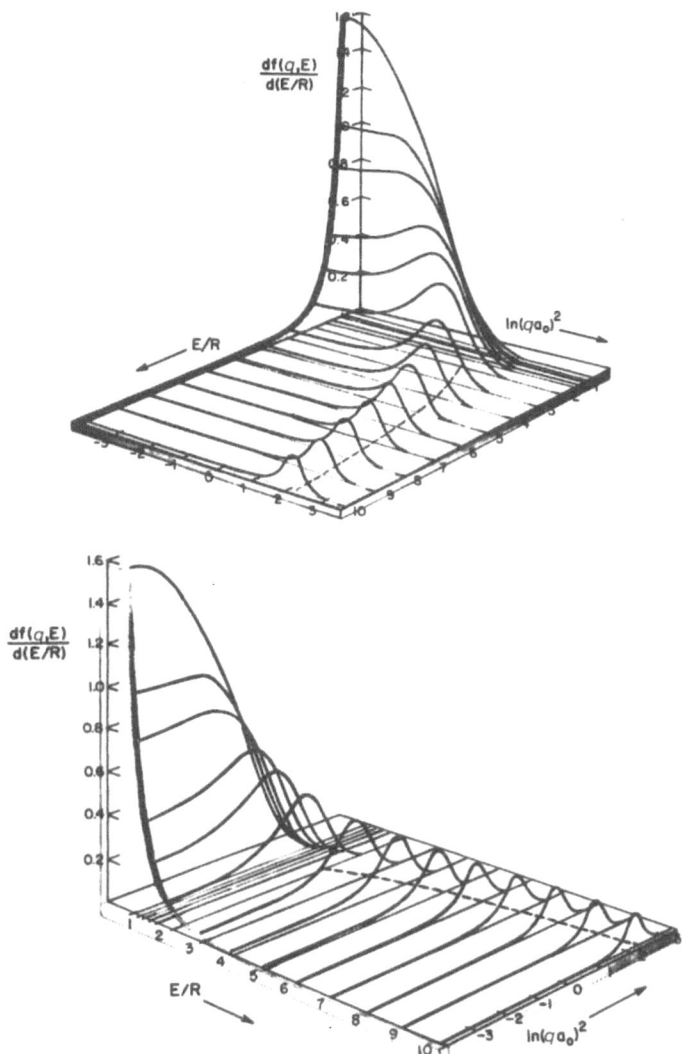

Fig. 2.3. Photographs of a plastic model of the Bethe surface for atomic hydrogen. The horizontal axes for E/R and $\ln (q\,a_0)^2$ define the base plane. The vertical axis represents $R df(q, E)/dE$. The fourteen plates are placed at $E/R = 3/4,8/9,1,5/4,3/2,2,3,4,5,6,7,8,9,10$. The first two plates represent the discrete spectrum, in which case the vertical scale corresponds to $\frac{1}{2} n^3 f_n(q)$, n being the principal quantum number. The dotted curve on the base plane shows the location $(q\,a_0)^2 = E/R$ of the Bethe ridge, which is the main feature for large E/R. From [2.7].

28

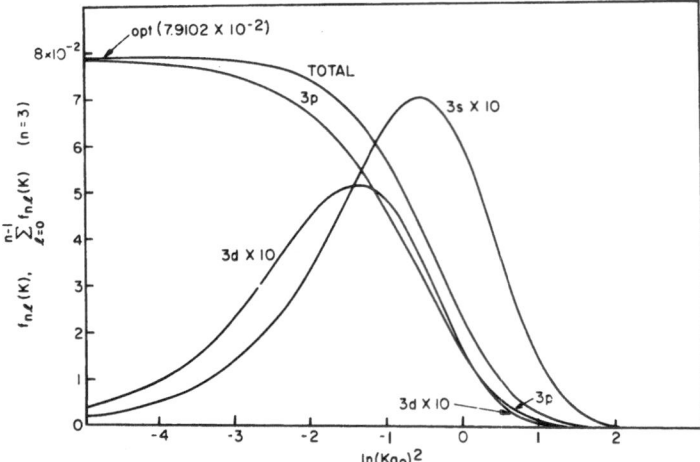

Fig. 2.4. Generalized oscillator strength for the transitions from the ground state to the $n = 3$ level of H. From [2.7].

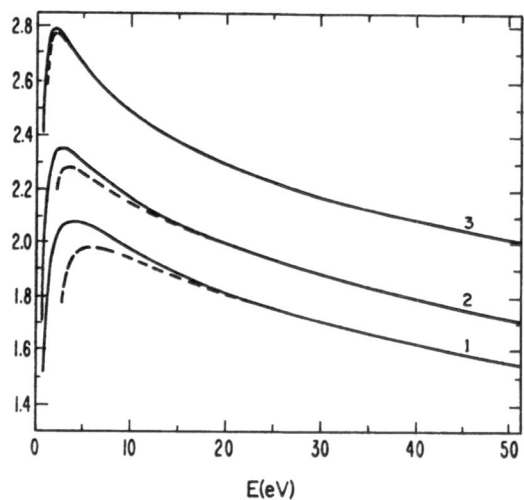

Fig. 2.5a. Total cross sections for the excitation of atomic hydrogen, calculated using the first Born approximation and the Bethe approximation. Curve 1: 3s→4p, Curve 2: 3p→4d, Curve 3: 3d→4f. Solid curves denote first Born approximation while dashed curves denote Bethe approximation. From [2.8].

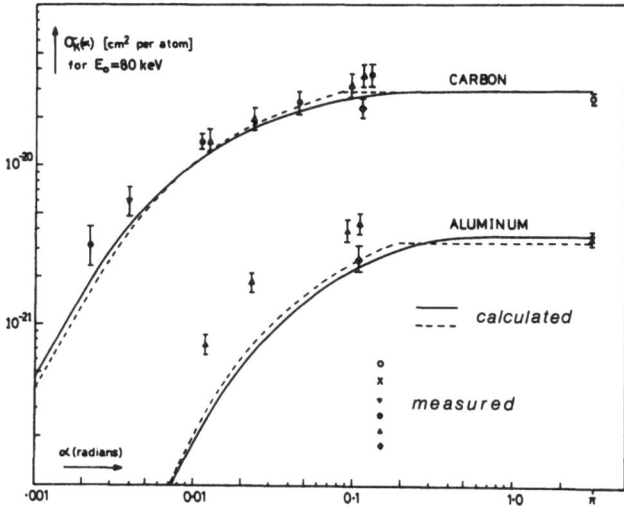

Fig. 2.5b. Integral and total cross-sections for K-shell ionization in carbon and aluminium, for an incident energy of 80 keV. From [2.10].

2.8. The Bethe Approximation

The scattering form factor S can be simplified by expanding e^{iqr} in Eq. (2.14). This so called Bethe approximation yields a convergent series as long as $qa_0 \ll 1$, since the hydrogenic wave functions are significant just for $r \le a_0$:

$$S_{\varphi'} = <\varphi'|\varphi> + iq <\varphi'|x|\varphi> - \frac{q^2}{2} <\varphi'|x^2|\varphi> - \ldots \quad (2.32)$$

The first term vanishes due to the orthogonality of φ, φ'. The next one contains a dipole matrix element which also governs optical transitions ($1s \rightarrow 3p$ in Fig. 2.4). For a particular excitation Eq. (2.32) is suited to estimate the order of magnitude of the cross section. In optically forbidden transitions ($s \rightarrow d$ or $s \rightarrow s$) the dipole matrix element vanishes, hence the form factor S and the cross section are dominated by the next important term $<\varphi'|x^2|\varphi>$ (see Fig. 2.4). Cross sections calculated by use of the Bethe approximation are shown in Fig. 2.5a for hydrogen [2.8].

2.9. Zonal Harmonics Expansion

For the solutions of the Schrödinger Eq. we make the *Ansatz*

$$\varphi_{nlm}(r, \Omega) = \frac{F_{nl}(r)}{r} Y_{lm}(\Omega). \tag{2.33}$$

Y_{lm} are spherical harmonics.

After separating out the angular part, we arrive at

$$\left[\frac{d^2}{dr^2} + E - \left(V(r) + \frac{l(l+1)}{r^2} \right) \right] F_{nl}(r) = 0 \tag{2.34}$$

for the radial part, where E is the energy of the atom electron (in Rydbergs) and r is given in units of the Bohr radius $a_0 = \hbar^2/me^2$. n, l are the usual energy and angular momentum quantum numbers, and V is the central potential, which is in general Coulombic $\propto -1/r^2$ only in the limits $r \to 0$, $r \to \infty$ or for hydrogen. For continuum states, n may be thought of as the (continuous) energy quantuum number of the electron. In order to calculate the form factor (2.14), the expansion of the plane wave

$$e^{i\vec{q}\vec{r}} = \sum_\lambda i^\lambda (2\lambda + 1) j_\lambda(qr) P_\lambda(\cos \vartheta), \tag{2.35}$$

where ϑ is the angle between $\vec{q}$ and $\vec{r}$, is used. j_λ are spherical Bessel functions, P_λ are Legendre polynomials. Assuming, for the moment, that only one atom electron participates in the inelastic interaction, and that the transition is to one definite final state $|\varphi' >$, the sums over b and i in Eq. (2.14) reduce to one term, and

$$S = \sum_\lambda i^\lambda (2\lambda + 1) < \varphi' | j_\lambda(qr) P_\lambda(\cos \vartheta) | \varphi > . \tag{2.36}$$

Insertion of the unity operator $\int r^2 dr \int d\Omega |r\Omega >< r\Omega|$ and use of Eq. (2.33) yield

$$S = \sum_\lambda i^\lambda (2\lambda + 1) \int_{4\pi} d\Omega Y_{l'm'} P_\lambda Y_{l'm'} \underbrace{\int_0^\infty dr \, F_{nl}(r) j_\lambda(qr) F_{n'l'}(r)}_{R^\lambda_{nln'l'}(q)} .$$

$$\tag{2.37}$$

The angular part of the integral is, due to the orthogonality of spherical harmonics, a fairly simple expression which selects a few terms of the sum over λ. The values of the integral are tabulated as functions of l, l', λ, m, m'. They are known as Wigner-3j-symbols and are related to the Clebsch-Gordon coefficients [2.11].

For calculation of $\partial^2\sigma/\partial E\partial\Omega$, one has to sum over all form factors S for the various final state angular momenta at definite n', and then calculate SS^* (cf. Eq. (2.14)). One can show that the cross terms in the product vanish [2.11] and that only the first few terms for small l' are important [2.13].

Eventually, S, GOS and $\partial^2\sigma/\partial E\partial\Omega$ can be expressed as a weighted sum over a few $R^\lambda_{nl,n'l'}(q)$ which have to be computed numerically in the non-hydrogenic case.

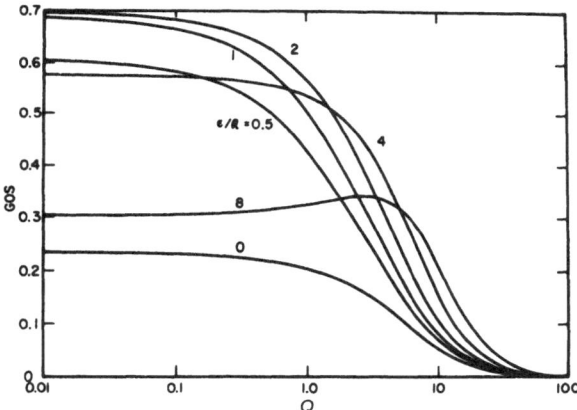

Fig. 2.6. Generalized oscillator strengths (GOS) per unit energy in Rydbergs for continuum transitions of the 2p shell of Al (binding energy 5.947 Ry). The curve parameter is the energy (in Rydbergs) above threshold. From [2.13].

Theoretical calculations of GOS and $\partial^2\sigma/\partial E\partial\Omega$ for many-electron atoms are prevented mainly by the lack of sufficiently accurate wave functions. Use of a self-consistent field (SCF) approach has proven feasible in this respect. The electrons are considered moving independently in a central SCF. We shall not concentrate on details here. SCF wave functions and effective potentials are tabulated

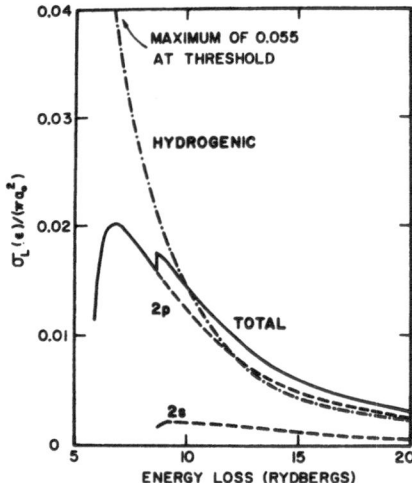

Fig. 2.7. Total L-shell continuum energy-loss cross section for 1-MeV protons. The solid curve has been obtained by the method described in the text, the dashed curves are the contributions of the individual subshells, and the dot-dashed curve is the hydrogenic result. From [2.13].

(e.g. [2.12]). As an example, the GOS thus obtained for continuum transitions from the $2p$-shell of Al are given in Fig. 2.6. Note the increase of the cross section with increasing energy loss, which is in contrast to the hydrogenic calculations [2.13]. See also Fig. 2.7.

2.10. Qualitative Interpretation

These results can be understood qualitatively as follows: Eq. (2.34) is formally equivalent with the Schrödinger Eq. for a one-dimensional system with effective potential $V(r) + l(l + 1)/r^2$ (see Fig. 2.8). The l-dependent part can be interpreted as a centrifugal barrier which prevents the wave function from penetrating the inner parts of the atom. Compare also Chapter 1!

The dominant terms in Eq. (2.37) come from the optically allowed transitions for Al $2p$: $l = 1 \rightarrow l' = l \pm 1$. Hence, $R^{\lambda}_{n1,n'2}$ and $R^{\lambda}_{n1n'0}$ are the decisive integrals. We shall investigate the case of small q, where $j_{\lambda} \approx const$. Then R can be estimated from the

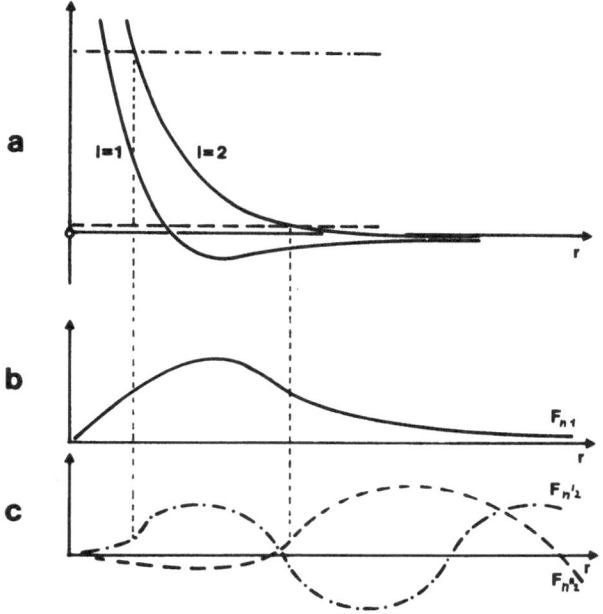

Fig. 2.8. a) Centrifugal barrier for angular momenta $l = 1$ and $l = 2$, b) Initial and c) final state wave functions for $p \to d$ transitions to the continuum (schematic). The final state wave function with higher energy $F_{n''2}$ overlaps F_{n1} more.

overlap of the initial and final state wave functions. Of course, the approximation $j_\lambda \approx const$ is useless for calculations—due to the orthogonality of $|\varphi >$ and $|\varphi' >$ the integral R would vanish—but it is still true that R is small as long as the overlapping parts of $|\varphi >, |\varphi' >$ are small.

Now, the electron in an initial state "sees" a centrifugal barrier $l = 1$, the continuum final state sees $l' = 2$. (The transition to $l' = 0$ contributes negligibly since the s-wave does not much overlap the p-wave). For small positive energies immediately above ionisation threshold there is practically no overlap of p- and d-states and $R^\lambda_{n1,n'2}$ is small. At higher final state energies the continuum wave function penetrates the atom more and more; consequently the overlap increases; and so does S. This effect causes what is called "delayed edges" in $L_{II,III}$-shell-transitions [2.14]. Eventually, GOS decreases due to destructive interference of the overlapping oscillatory functions $F_{nl}, F_{n'l'}, j_\lambda$.

2.11. The Total Elastic Cross Section

The scattering probability amplitude can be written as

$$f(\vartheta) = \frac{1}{2ik} \sum_{l=0}^{\infty} (2l+1) P_l(\cos\vartheta)(e^{2i\delta_l} - 1) \qquad (2.38)$$

where P_l are Legendre-polynomials, k is the wave number of the incident wave, and δ_l are partial wave phase shifts [2.11]. In order to obtain the total cross section, integrate $|f(\vartheta)|^2$ with respect to Ω:

$$\sigma_t = 2\pi \int_0^\pi |f(\vartheta)|^2 \sin\vartheta d\vartheta$$

$$= \frac{2\pi}{4k^2} \sum_{l,l'} (2l+1)(2l'+1)$$

$$(e^{2i\delta_l} - 1)(e^{-2i\delta_{l'}^*} - 1) \cdot \int P_l P_{l'} \sin\vartheta d\vartheta$$

$$= \frac{4\pi}{4k^2} \sum_l (2l+1) |(e^{2i\delta_l} - 1)|^2 \qquad (2.39a)$$

$$= \frac{\pi}{k^2} \sum_{l=0}^{\infty} (2l+1) \cdot |2i \sin\delta_l e^{i\delta_l}|^2$$

$$= \frac{4\pi}{k^2} \sum_{l=0}^{\infty} \sin^2\delta_l (2l+1)$$

where we have used the orthogonality of P_l and the identity $e^{2i\delta} - 1 = 2i \sin\delta e^{i\delta}$. We note in passing that comparison of (2.39) and (2.38) for $\vartheta = 0$ yields the "optical theorem"

$$\sigma_t = \frac{4\pi}{k} \cdot Im(f(0)) \qquad (2.39b)$$

From Eq. (2.38), it is seen that the problem of calculating the scattering amplitude $f(\vartheta)$ is equivalent to determination of δ_l, which are the phase shifts of the partial waves $j_l(kr)$ in the expansion Eq. (2.35) of the incident plane wave of wave vector $\vec{k}$

$$e^{i\vec{k}\vec{r}} = \sum_{l=0}^{\infty} (2l+1)i^l j_e(kr)P_e(\cos\vartheta) \tag{2.35}$$

where ϑ is the angle between $\vec{k}$ and $\vec{r}$.

Since

$$\lim_{r\to\infty} j_l(kr) = \frac{\sin(kr - \frac{l\pi}{2})}{r} \tag{2.40}$$

we have for the *scattered* partial wave j_l' with shift δ_l the asymptotic form

$$j_l' \doteq \frac{1}{r}\sin(kr - \frac{l\pi}{2} + \delta_l). \tag{2.41}$$

Calculation of phase shifts δ_l from a given potential is, in general, a formidable task, which is, in the limiting cases of high and low energy scattering, definitely simplified. We assume for both limits that the potential of the scatterer is significant within a sphere of radius a and can be neglected outside (think of a hard sphere).

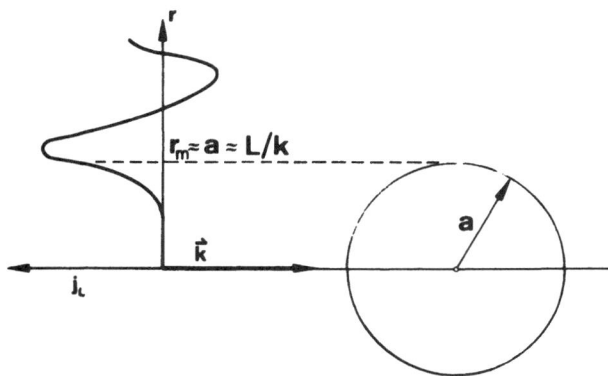

Fig. 2.9. Radial part of partial wave with angular momentum L. Partial waves with $l > L$ are negligible inside the interaction sphere of radius a.

When the interaction of the incident wave and the scatterer is confined to a sphere of radius a, only those partial waves experience a phase shift which do not vanish inside the sphere $r = a$.

The maximum of $\dot{\jmath}_l$ is approximately located at $kr_m = \sqrt{l(l+1)}$. for smaller r, $\dot{\jmath}_l$ decreases rapidly (e.g. [2.17]) as sketched in Fig. 2.9.

It follows that only those partial waves are significantly shifted for which $l \le ka$. In practice, one may truncate the series (2.38) at

$$L \approx ka. \tag{2.42}$$

Two limiting cases are of particular interest:

a) high energy scattering: When k is very large ($k \gg 1/a$) then many partial waves contribute to (2.39). One can replace the sum over l by $\int dl$. In a crude approximation, the phase shifts δ_l are assumed to vary randomly with l, so the factor $\sin^2 \delta_l$ in the integral can be set equal to its mean value:

$$\frac{1}{2\pi} \int\limits_0^{2\pi} \sin^2 x\, dx = \frac{\pi}{2\pi} = \frac{1}{2}. \tag{2.43}$$

Hence, the total scattering cross section is, according to (2.39)

$$\sigma_{t,h} \doteq \frac{4\pi}{k^2} \int\limits_0^{L=ka} (2l+1) \sin^2 \delta_l\, dl \doteq \frac{4\pi}{k^2} \frac{(ka)^2}{2} = 2\pi a^2. \tag{2.44}$$

For high energy incident particles, the quantum mechanical cross section is *twice* the classical value;

b) low energy scattering: When the momentum of the incident particle is very small ($k \ll 1/a$), the condition (2.42) for truncation of the series Eq. (2.39a) for σ_t is $L \ll 1$. Only the partial wave $\dot{\jmath}_0(kr)$ with angular momentum $l = 0$—the s-wave—will be shifted. From Eq. (2.39), we have for the low energy case

$$\sigma_{t,l} \doteq \frac{4\pi}{k^2} \cdot \sin^2 \delta_0. \tag{2.45}$$

δ_0 can be calculated from the scattering potential. A hard sphere of radius a, for instance, shifts an emerging spherical wave from the origin to $r = a$. The phase shift is then

$$\delta_0 = 2\pi \cdot \frac{a}{\lambda} = k \cdot a \ll 1 \tag{2.46}$$

and Eq. (2.45) is

$$\sigma_{t,l} \doteq 4\pi a^2 \tag{2.47}$$

which is the *fourfold* classical cross section.

Our result Eq. (2.44) seems to have general consequences since we did not assume microscopic objects in the derivation. It holds for macroscopic objects as well, such as billiard balls. Does quantum mechanics, then, influence the strategy of a player? When the cross section is twice the geometrical cross section, even an ill-trained player, like me should have a good score.

The puzzle is solved as follows: In Chapter 1, we derived an expression for the angular range within which the classical model no longer applies. We conclude that the increase in scattering cross section as compared to the classical one occurs within this critical angle (see Fig. 2.10).

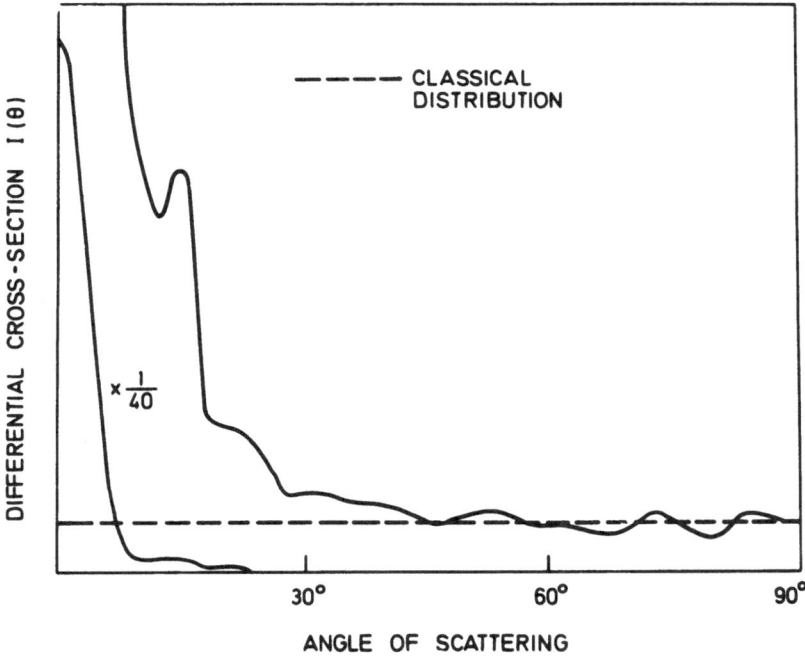

Fig. 2.10. Classical and quantum mechanical differential scattering cross sections for a hard sphere. From [1.4].

A billiard ball moves at approximately $10^{-1} m/s$ and has a mass of, say, $10^{-1} kg$. Hence, $\lambda = 2\pi/k = 2\pi\hbar/mv \sim 10^{-31} m$. The critical angle ϑ_c according to Eq. (1.29) is, for billiard balls,

$\vartheta_c \sim 10^{-31} rad$. Deviations from classical mechanics occur only within a scattering angle of $10^{-31} rad$ which is so close to zero that in practice one cannot discern those cases from collisionless trajectories. Paradoxically, the increased quantum mechanical cross section bears no practical consequences for macroscopic objects.

2.12. The Ramsauer-Townsend Effect

As discussed in Chapter 1, in 1921 Ramsauer and Townsend independently discovered that the scattering cross section for slow electrons on the rare gases Ar, Xe, Kr decreased to a minimum for a primary electron energy of 0.7 eV [1.5]. This was in contradiction to classical scattering theory which demands that $\partial^2 \sigma / \partial \vec{\Omega}$ decreases monotonously with electron velocity.

An explanation can be given in terms of partial wave phase shifts δ_l which relate to the scattering probability amplitude as [2.11]

$$f(\vartheta) = \frac{1}{2ik} \sum_{l=0}^{\infty} (2l+1)(e^{2i\delta_l} - 1) P_l(\cos \vartheta). \qquad (2.38)$$

Since scattering on a center of force conserves the angular momentum (with respect to that center), the scattering matrix **S** commutes with the angular momentum operator. It follows that the angular momentum eigenstates $|\psi_l >$ are likewise eigenstates of **S**. Since **S** is unitary, we have

$$\mathbf{S}|\varphi_l >= S_l|\varphi_l >= e^{2i\delta_l}|\varphi_l > . \qquad (2.48)$$

Due to the unitarity of **S** the exponential has modulus 1. It can be written as $e^{2i\delta_l}$ [2.11]. Note that the same factor appears in Eq. (2.38). Because of (2.48) it is useful to expand the wave function $|\psi_I >$ of the incoming electron as

$$|\varphi_I >= \sum_{l=0}^{\infty} a_l|\varphi_l > . \qquad (2.49)$$

The scattered electron is

$$|\varphi_S> = \mathbf{S}|\varphi_I> = \sum_{l=0}^{\infty} a_l S_l|\varphi_l> = \sum_{l=0}^{\infty} a_l e^{2i\delta_l}|\varphi_l> . \qquad (2.50)$$

The important point is that the expansion (2.49) may, to a good approximation, be truncated at $L = q \cdot R$ (where R is the effective radius of interaction) so as to describe the incoming electron within that radius (cf. Fig. 2.9).

At the Ramsauer-Townsend minimum, $q \approx 5.10^9 m^{-1}$, and R is of atomic scale, $R \approx 10^{-10}m$, hence $L \approx 0.5$. Consequently, only $|\psi_0>$ will be considerably scattered according to (2.50). Higher order partial wave shifts δ are likely to be small. Now, suppose the phase shift δ_0 behaves as shown in Fig. 2.11. At velocity v_0, we have $\delta_0 = 3\pi$. Use of Eq. (2.39) then shows that $\sigma_T \approx 0$. The scattered wave is nearly identical with the incoming one which means that the atom is transparent for the electron beam. Inspection of Eq. (2.38) tells that $f(\vartheta)$ vanishes under these circumstances, too.

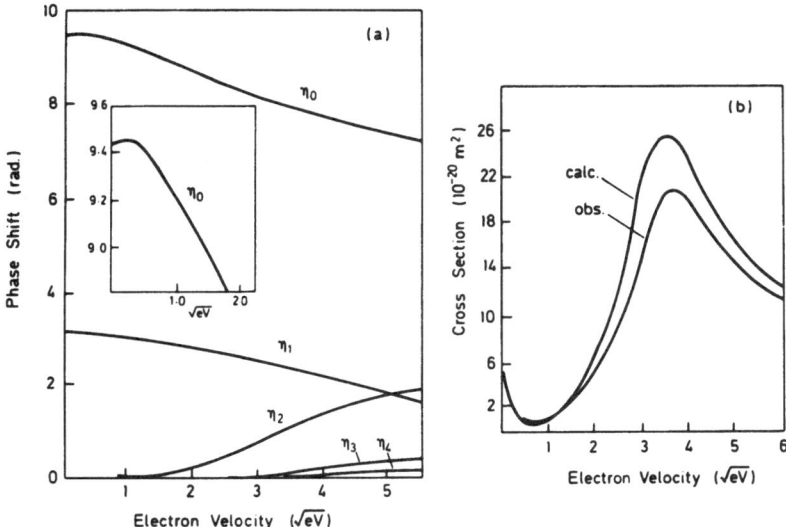

Fig. 2.11. a) Phase shifts $\eta_0, \eta_1, \eta_2, \eta_3, \eta_4$ calculated theoretically for argon as a function of electron velocity.
b) Comparison of the cross-section calculated using these phase shifts with those observed. From [1.4].

In Fig. 2.11 calculated phase-shifts $\delta_0 - \delta_4$ are drawn for Ar. It is obvious that the Ramsauer-Townsend effect is indeed due to the behaviour of δ_0. The higher order phase shifts become effective only at higher electron velocities, indicating that at sufficiently low velocity scattering does not influence the partial waves with $l > 0$.

This chapter should be considered nothing other than a short glimpse of the scattering problem. There is a variety of much more elaborate approaches than those discussed— such as second order perturbation theory or accounting for exchange contributions, or semiclassical approaches— each adequate for particular cases, but none of them generally applicable. Although the theory of inelastic collisions of electrons with atoms was originated by H. Bethe as early as 1930 almost in its final form [2.1], we are far from a general understanding of these processes regarding both theory and experiment [2.7]. For recent examples, see for instance [2.15], [2.16].

3. Practical Aspects of Absorption Edge Spectrometry

3.1. A Survey of Applications

Since each element has ionisation edges at characteristic energies, the main aspect in edge spectroscopy is elemental analysis. The task of deciding which element(s) a particular sample consists of, is relatively easy to perform by observation of structure at ionisation energy losses in the elsewhere smooth spectrum. Quantitation of loss spectra is quite another story, complicated by such facts as insufficient knowledge of cross sections, instrumental aberrations and instabilities, or masking effects (background intensity, multiple scattering, etc.). At present an accuracy of 20 % in quantitation seems to be realistic, although better figures have been obtained in particular cases [3.1], [2.19]. The same holds for the lower detection limit of mass fraction which is, for routine application, on the order of 5 atom % [3.18], whereas under special circumstances, minima of some 0.01 at % [3.10] to 0.5 at % [3.2] are given. As to the absolute lower detection limit, there are several calculations, yielding some ten atoms [3.19], [3.9] up to some 100 atoms [3.20].

A more complicated problem is investigation of the edge shape. In principle, the shape of an absorption edge contains information on the ground state wave function, as discussed in the foregoing chapter. The question arises as to whether the wave function can be retrieved from a precise measurement of the edge structure. It turns out that the problem is ill-posed in the sense that different wave functions may yield the same spectral feature. At the root of this ambiguity is the phase problem, well known in diffraction physics, which prevents retrieval of the phase of a complex quantity from measurement of its modulus.

Compton scattering offers a close approach to the retrieval of wave functions. From Compton scattering of γ-rays on gas atoms it has been known for a long time that the Compton profile mirrors the momentum distribution of the scatterer (= the core electron) in its ground state [3.3]. It can be shown that this information delivers the autocorrelation of the wave function, which is indeed less than the wave function itself, but more than charge density autocorrelation functions which usual diffraction techniques may yield.

Only recently has this technique been applied on electrons as a probe, then termed ECOS (electron Compton scattering [3.4]). Speaking rigorously, one measures ionisation probabilities along a cut through the Bethe surface at constant momentum transfer. First reports on ECOS experiments on solids seem to confirm its power, especially with respect to the investigation of bonding in molecules [3.5]. We shall give a brief account on ECOS in the following.

There is another technique only recently contrived [3.6], although the underlying effect has been well known in dynamical diffraction theory as "channeling" for some decades: By properly orienting a single crystal with respect to the incident plane wave, charge density maxima can be set up either at the lattice planes or in between. (So to speak, the probe electron behaves as if confined to one of two channels). A combination of the channeling phenomenon with edge spectroscopy offers site-specific elemental analysis, which might be especially important for the investigation of impurity locations.

Channeling of electrons is caused by the regular arrangement of scattering atoms in a lattice. Another effect caused by atomic periodicity is the structure at the far energy side of an ionisation edge, known as EXELFS (extended energy loss fine structure). Whereas in channeling the *probing* electron sees the regular arrangement of scatterers, it is the *ejected* electron in EXELFS. The fine structure at the edge's slope is caused by destructive and constructive interference of wave functions scattered from neighbouring atoms, thus providing a means of probing nearest neighbour distances [3.7]. In the seventies, EXELFS seemed to become a powerful counterpart to EXAFS (extended X-ray absorption fine structure). However, development of intense synchrotron radiation sources elicited work on EXAFS, whereas data acquisition problems in EXELFS made the promise unclaimed in its original form [3.21]. ELNES (Energy Loss Near Edge Structure) is a related matter, the difference to EXAFS being the smaller energy range inspected beyond the edge.

ELNES is generally attributed to transitions to unoccupied conduction states or molecular levels [3.15]. As we shall see, the same applies to EXAFS, from a slightly different viewpoint. So, it is superfluous to discuss both of them separately. It should be noted, however, that the whole subject is too complex to be apt to a rigid interpretation [3.16].

3.2. Microanalysis

The differential cross section $\partial^2\sigma/\partial E\partial\Omega$ is related to the detected probe current Δj within $\Delta E \Delta \Omega$ according to Eq. (2.3) as

$$\Delta j = j_i \int\limits_{\Delta E, \Delta\Omega} dE d\Omega \frac{\partial^2\sigma}{\partial E \partial \Omega} = j_i \Delta\sigma \qquad (3.1)$$

where j_i is the incident current density. As a cross section, $\Delta\sigma$ has dimension $m^2/$atom. Hence, a specimen with number density of atoms n, thickness d and area F causes, within $\Delta E \Delta \Omega$ a probe current ΔJ (see Fig. 3.1).

$$\Delta J = nFd\Delta\sigma j_i. \qquad (3.2)$$

A quantity measurable without standards or calibration in an energy loss experiment is the ratio of incident and scattered current densities, which is, for small scattering angles,

$$R(\Delta E, \Delta\Omega) = \frac{\Delta J/F}{j_i} = nd\Delta\sigma(\Delta E, \Delta\Omega). \qquad (3.3)$$

For "pure" specimens of known density n Eq. (3.3) can be used to either obtain d (which we shall not dwell on further) or $\Delta\sigma$, which is meaningful in two respects:

a) $\Delta\sigma$ can be compared with theoretical calculations of the cross section—this is important since the theory of cross sections is far from being settled, as mentioned in Chapter 2.

We note in this context that, for practical purposes, $\Delta\sigma$ is conveniently calculated in the hydrogenic approach. Computer programs are available for K- and L-shell losses [2.10], [3.8].

b)   For microanalysis, $\Delta\sigma$ can be obtained from standard samples to calibrate results of measurements on specimens of unknown composition.

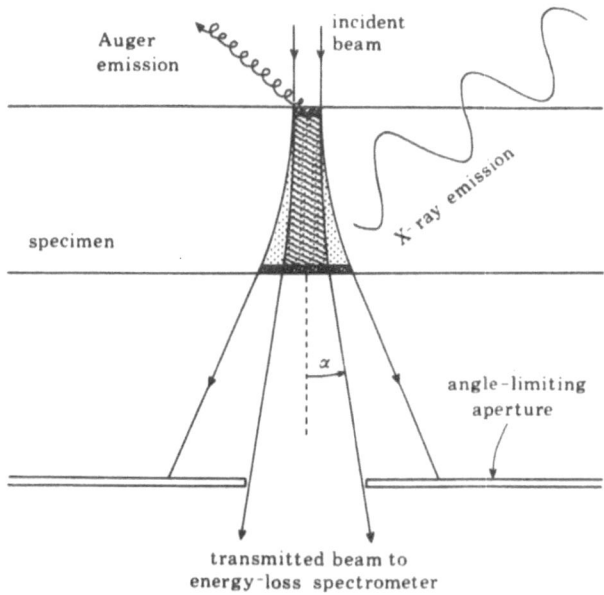

Fig. 3.1. Transmission of an electron beam through a thin sample showing X-ray and Auger electron emission due to an inelastic scattering event. $\Delta\Omega$, and hence $\Delta\sigma$, depends on the aperture-angle α. From [3.22].

Sometimes $\Delta\sigma$ is expressed in units of the total cross section for ionisation from a particular shell i

$$\sigma_i = \int\limits_{E_i}^{\infty} dE \int\limits_{4\pi} d\Omega \frac{\partial^2 \sigma}{\partial E \partial \Omega} \tag{3.4}$$

then referred to as collection efficiency $\eta = \Delta\sigma/\sigma_i$ (see Fig. 3.2).

It can be seen from Fig. 3.2 that η_K for fixed $\Delta E, \Delta \Omega$ decreases with increasing atomic number Z. This is due to the fact that the angular halfwidth of the edge profile increases with Z, hence less intensity is detected within fixed $\Delta \Omega$ for higher Z. The same figure shows the fluorescent yield as a function of Z (dashed line). The supplement to 1 is the Auger-yield (An atom which has been ionized by the probing electron either emits an X-ray or an Auger electron.)

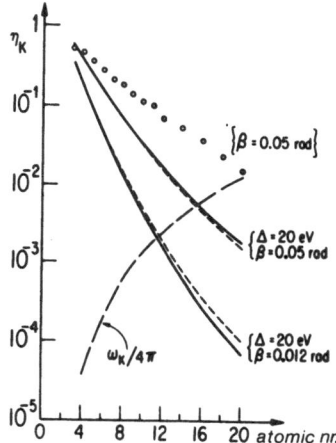

Fig. 3.2. Collection efficiency, η_K for electrons which have lost energy in producing a K-shell excitation or ionization. The solid curves have been calculated assuming a Lorentzian angular distribution of the differential cross section. The energy window $\Delta = 20$ eV, β is the collection angle. The open circles are calculated approximately with a variable energy window given by $\Delta = 1/2[E_K(Z) - E_K(Z-1)]$. The incident electron energy is assumed to be 25 keV. The long-dashed line is the efficiency for detecting X-rays assuming the X-ray detector counts all X-rays impinging upon it and that it subtends $1/4\pi$ steradian at the specimen. From [3.10].

Eq. (3.2) answers the question of what the theoretical lower detection limit for an element is when for ΔJ the minimum detectable current ΔJ_{min} which is accessible experimentally is inserted:

$$N_{min} = \frac{\Delta J_{min}}{\Delta \sigma j_i}. \tag{3.5}$$

See Fig. 3.3, where the minimum detectable mass is plotted.

A comparison with the X-ray minimum detectable mass shows that practically ELS is superior up to $Z \approx 20$.

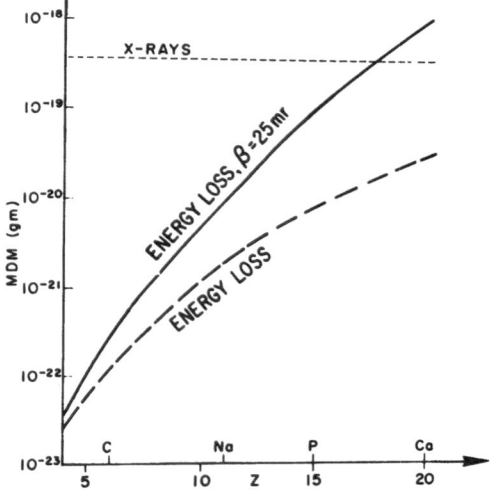

Fig. 3.3. Minimum detectable mass as a function of atomic number Z for 25 keV electrons. An energy window $\Delta = 1/2[E_K(Z) - E_K(Z-1)]$, a current density $J = 10^2 A/cm^2$ and a minimum practical counting rate of $R_M = 10^2$ counts/sec have been assumed for calculations. The solid line pertains to a spectrometer aperture half-angle of $\alpha = 25$ mrad while the dashed line is for a variable spectrometer aperture chosen so that $\eta_\alpha \approx 1$. The upper dashed curve is for X-ray detection and a 100 % efficient detector which subtends $1/4\pi$ steradians at the specimen. From [3.10].

This is because

a) the X-ray fluorescent yield is smaller than 1 because of the additional occurrence of Auger electrons in recombinant processes, and

b) the fluorescent yield is proportional to the X-ray energy.

From Fig. 3.3 one is tempted to infer that with optimum conditions as few as 10 atoms of carbon are detectable. However, Eq. (3.5) is a theoretical lower bound neglecting background intensity, abundance of other elements, counting statistics and so on.

Depending on which effects are considered in calculations, and which counting rates and probe currents are accepted, there are various results. We present two of them (Fig. 3.4), one for an iron matrix 50 nm thick at 100 kV primary energy and a current density of 10 A/cm^2 [3.9]; the other one for a carbon matrix 20 nm thick, at 25 kV and 10 A/cm^2 [3.10].

In practice, often even these minima are not obtained due to

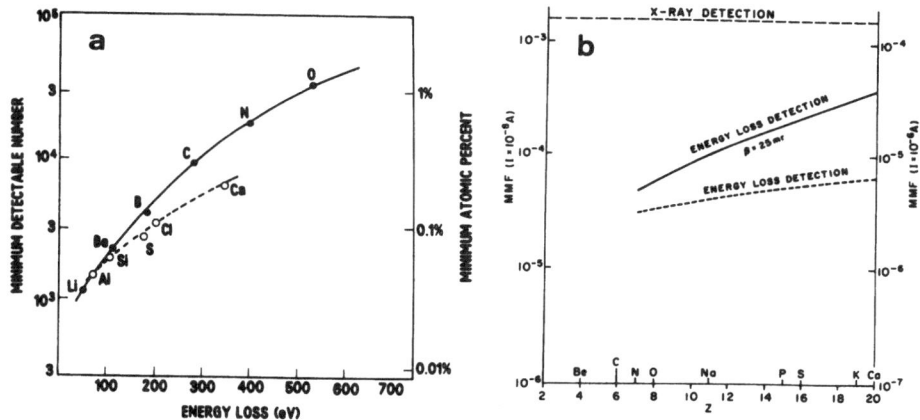

Fig. 3.4. a) The minimum detectable number of atoms of various elements from Li to Fe in a 50 nm thick iron foil, and the equivalent data plotted as an atomic percentage. $J_i = 10^{-10} A$ at 100 kV. From [3.9].

b) The minimum mass fraction (MMF) of various elements in a 20 nm thick carbon matrix. The vertical scales correspond to beam currents of $J_i = 10^{-8} A$ (left scale) and $J_i = 10^{-6} A$ (right scale), both at 25 kV. The short-dashed curve corresponds to a variable spectrometer aperture which is large enough so that $\eta \approx 1$. Also shown are results for X-ray detection with a 100 % efficient detector which subtends $1/4\pi$ steradian at the specimen. From [3.10].

various experimental instabilities, such as drift effects, contamination or damage of the specimen in the beam. Recently, however, detection of 10 atoms uranium has been reported [3.2].

3.3. Electron Compton Scattering

In classical scattering theory, the differential cross section for scattering on a target electron at rest is a Dirac δ-function defined so that energy and momentum of the interacting particles be conserved. Quantum mechanically, the δ-function is broadened to

what is called the "Bethe ridge". The broadening is essentially due
to the momentum uncertainty of the target electron in its ground
state.

Classically, the same effect is expected for an ensemble of atoms.
The initial momenta of their bound electrons are distributed around
the mean value, thus imparting an additional Doppler shift to the
scattered particle. The resulting ensemble scattering cross section
for fixed scattering angle is known as the "Compton profile". Obviously, the Compton profile contains information on the momentum
distribution of the target electrons in the ensemble. In fact, the
measured profile $P(q)$ is a projection of the initial momentum distribution of the target electrons onto the direction of scattering
[3.3], [3.11].

Since,

$$P(q) = \Psi(q)\Psi^*(q) \tag{3.6}$$

where $\Psi(q)$ is the wave function in momentum space, we have direct access to the modulus of the wave function. The probability
distribution $\tilde{P}(r)$ is obtained by a Fourier transform of $P(q)$. The
Fourier transform of a product yields a convolution (see Chapter
4), hence

$$\tilde{P}(r) = \int e^{iqr} P(q) dq = \int \tilde{\Psi}(r')\tilde{\Psi}^*(r - r')dr'. \tag{3.7}$$

So, the Fourier transform of the Compton profile yields the self-
convolution of the ground state *wave function*. This is an important
result as compared to diffraction experiments which provide only
the *charge density* autocorrelation (for definition cf. Eq. (2.18)).
Contrary to the latter, which is always positive, $P(r)$ exhibits also
negative parts and thus allows for a partial retrieval of phase information.

As an example, Fig. 3.5 shows $\tilde{P}(r)$ for two crystallographic
directions in Beryllium metal together with results of model calculations. Note the excellent agreement between theory and experiment. Important applications of Compton scattering are the
investigation of bonding in molecules [3.5] and anisotropies of the
wave function in crystals [3.12].

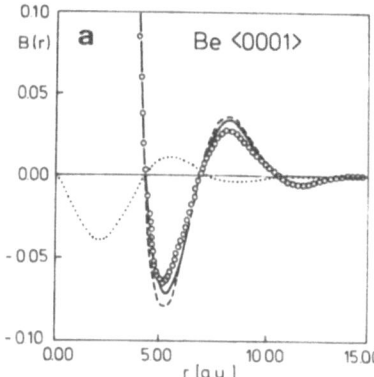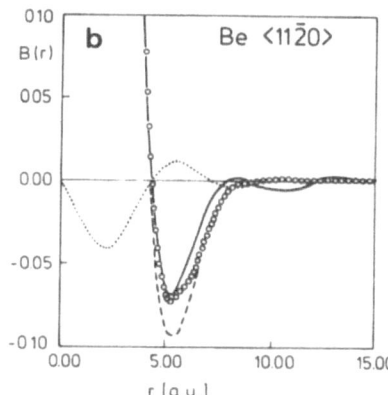

Fig. 3.5. a) Fourier transformed <0001> Compton profiles of Beryllium: experiment (circles), Hartree-Fock calculations (continuous curve), density-functional calculations (dashed curve), correlation correction (dotted curve). b) As Fig. 3.5a), but <11$\bar{2}$0> direction. From [3.12].

The technique of Compton scattering has been designed originally for γ-rays as a probe. Experimenters have to face various difficulties, such as moderate momentum resolution, long counting times (on the order of days), or the need for large samples ($\sim cm^2$). Only recently has using electrons as a probe for solids [3.4] been tried, the technique then designated ECOS(S) which is an acronym for electron Compton scattering (from solids). As it stands, at least two of the difficulties with γ-ray probes are not encountered in ECOS(S), viz. large samples and extremely long counting times. Since experiments are done with electron microscopes small samples can be used. Counting times seem to range within hours instead of days [3.4] (due to the higher cross section for electron scattering, see Chapter 2). However, whether ECOS(S) is useful for standard application will be seen only after more extensive basic studies have been presented.

3.4. Site-specific Excitations

As is well known from electron diffraction theory, the probing
electrons can be channeled in a single crystal which is properly
oriented. When just two diffraction spots are equally well excited
(two beam case) the wave field in the crystal is set up of two equally
strong types of Bloch waves, one of which has maxima at the lattice
points, the other in between [3.13].

By slightly tilting the crystal out of the Bragg position, one of
these two Bloch waves is enhanced, whereas the other one is atten-
nuated. Thus by tilting the crystal, the specimen can be probed at
either the lattice points or in between. Fig. 3.6 shows calculated
current distributions across two unit cells in Al. The figure is a cut
perpendicular to the (111)-planes [3.14]. Fig. 3.7 is a scheme of the
beam geometry, two orientations relative to the reciprocal lattice
are marked.

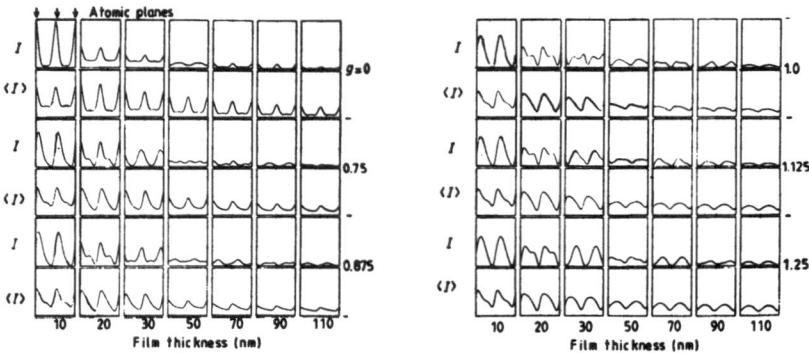

Fig. 3.6. Calculated electron current distributions, I, across two unit cells at a
variety of depths into an aluminium foil and depth averaged, $< I >$, for tilts of
$g = 0$ (Laue condition); 0.75, 0.875, 1 (Bragg condition), 1.125 and 1.25 using
an absorption ratio of 0.04 for 100 kV incident electrons. From [3.14].

Fig. 3.8 is an example of site-specific absorption in chromite
spinel [3.6]. The relative strength of Cr- and Fe-edges is clearly
different for the two crystal orientations depicted on the left-hand
side. The geometry of Fig. 3.8a corresponds to probing nearly *at*
the (400)-lattice planes (strong absorption inside the (400)-Kikuchi-
band), whereas the geometry of Fig. 3.8b preferentially probes

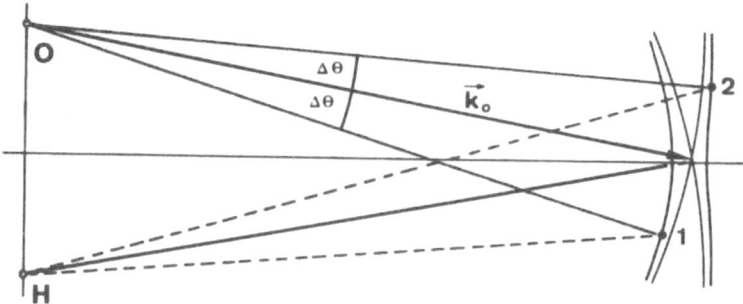

Fig. 3.7. Beam geometry with respect to reciprocal lattice points $0, H$, to be set up for the two-beam case in the electron microscope in order to excite predominantly Bloch waves "1" or "2". The latter have maxima *at* the $(H00)$-lattice planes. $\Delta\Theta$ is the tilt angle off the exact Bragg position.

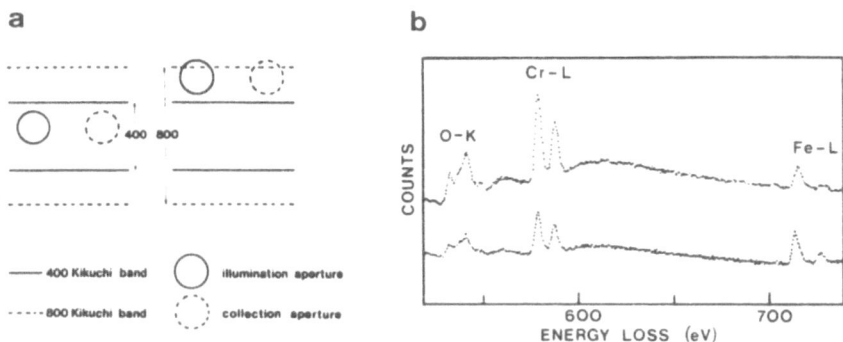

Fig. 3.8. a) Diffraction geometries used to set up standing-wave fields with maxima at different sites. Left: Octahedral sites selected; right: tetrahedral sites selected. Note that the crystal is in ($H\,00$) Bragg reflection condition when the apertures are exactly at the corresponding ($H\,00$) Kikuchi line.
b) Electron energy-loss spectra of chromite spinel taken under the two conditions illustrated in a). From [3.6].

between the (400)-planes. Obviously, there are more iron atoms located at these latter positions.

Since this technique is rather new and complicated (correct alignment of probe and target, thin single crystals) it has not been used extensively thus far.

3.5. Extended Energy Loss Fine Structure (EXELFS)

This effect is similar to the fine structure superimposed on the far side of an X-ray absorption edge (EXAFS) (Fig. 3.9).

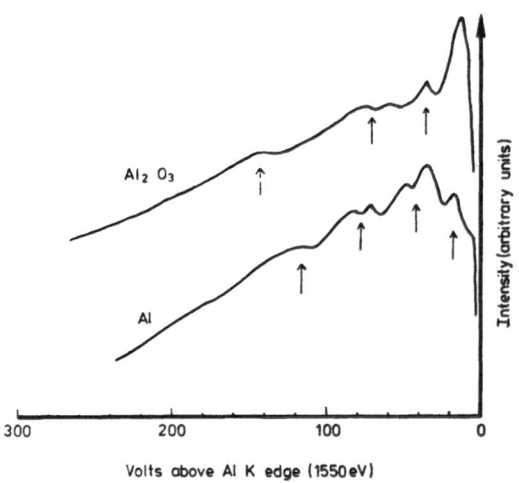

Fig. 3.9. EXELFS in Al and Al_2O_3 (50 nm thick), indicating nearest-neighbour maxima with arrows. From [3.7].

The structure is caused by interference of the wave function of the ejected electron with waves scattered from neighbouring atoms (Fig. 3.10).

At particular values of wave number Δk of the ejected electron, interference is constructive. An approximate formula for the resulting intensity is [3.7]

$$I(\Delta k) = |S(\Delta k)|^2 \Delta k \sum_j \frac{n_j e^{-\gamma r_j}}{r_j^2} \sin(2\Delta k r_j) \qquad (3.8)$$

where I is the intensity, S the atomic scattering factor, n_j the number of $j-th$ neighbours, r_j their distance, $e^{-\gamma r_j}$ a screening factor and the relation between Δk and the energy E above the edge is $E = \hbar^2 \Delta k^2 / 2m$. Since the wave number Δk of the ejected electron increases with the energy distance from ionization threshold (see

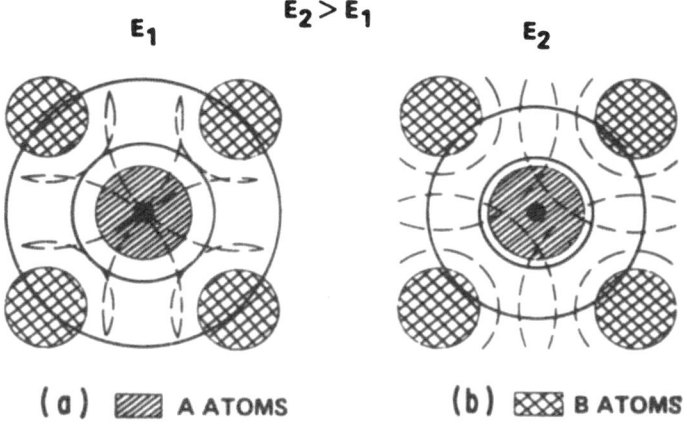

Fig. 3.10. Schematic representation of an EXELFS event for different energies E above threshold. The excited electronic state is centered about the A atom. The solid circles represent the crests of the outgoing part of the wave function. The surrounding B atoms diffract the outgoing part as shown by the dashed circles. Constructive interference is represented in (a) and destructive interference in (b). From [3.23].

Sect. 2.6), interference effects show up in energy. An alternative explanation can be formulated in terms of the density of states. According to Eq. (2.14), the differential cross section contains a summation over final states of the atom electron. The δ-function acts so as to select all states on a fixed "energy shell". For single atoms, the final state of the ejected electron is a free particle, the corresponding density of states reveals the well-known monotonous increase shown on the left of Fig. 3.11. The increase is compensated by the prefactor $1/q^4$ rendering the cross section decreasing monotonously with energy.

In the crystal, the density of states is no more monotonous because of its band structure. For bound electrons, there are gaps at particular energies, but even electrons in the continuum "feel" the periodic potential which causes more or less periodic variations in the density of states. These variations influence the scattering cross section— so, in fact, EXELFS are variations in the final density of states induced by the crystal potential. There is no principal difference to ELNES mentioned in Sect. 3.1, concerning the underlying effects.

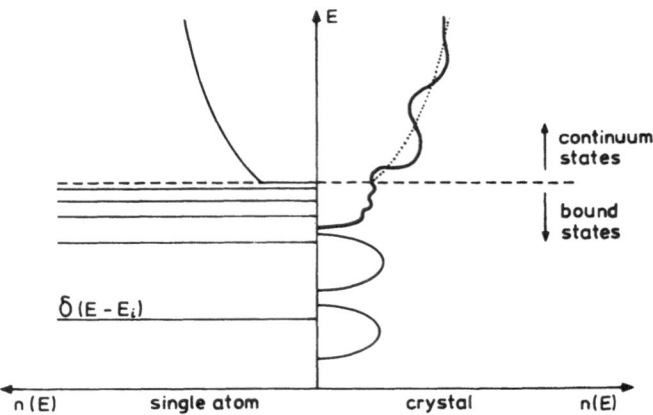

Fig. 3.11. Density of bound and continuum states for an electron of a single atom (left) and an electron in the periodic potential of the crystal ion cores.

From Eq. (3.8), r_j can be calculated for the first few coordination shells. However, due to a number of deficiencies such as very faint intensity variation and side effects from other edges, multiple scattering and edge energy dispersion, EXELFS interpretation is still a formidable task. A quantitative discussion of EXELFS has to include a number of effects, the most important of which seem to be multiple scattering of the ejected electron and the influence of the core-hole generated by the excitation (the core exciton). See, for example, [3.27].

The reader interested in more details of application is referred to the respective quotations in the text. For introductions and reviews the papers of Joy and Maher [3.1b], [3.9], Leapman [3.15], Colliex and Trebbia [3.18], Egerton [3.24], Silcox [3.25] and Johnson [3.26] are useful. The book of Egerton [3.28] is also highly recommended as a comprehensive review text on ELS applications.

4. Electrodynamics in Homogeneous, Isotropic Media

This chapter is devoted to the discussion of electromagnetic properties of homogeneous (as such they are unbounded in space) isotropic media, based on classical Maxwell theory. As we shall see, the central concept is that of the dielectric function (permittivity) ε. It determines the eigenmodes of the system as well as the medium's interaction with external perturbations, which is, in the context of inelastic scattering, a (fast) electron.

Unless explicitly stated, we shall not assume that the medium is conductive. On the other hand, the classical approach presented in the sequel is most useful for metals, since in this case inhomogeneities, anisotropy and whatsoever are of minor importance for a basic understanding of electromagnetic (frequency-dependent) properties. So, it is not by accident that the determination of the dielectric function $\varepsilon(k, \omega)$ from inelastic scattering experiments is predominantly carried out for metals. Such details as energetic location of electronic transitions or deviation from the ideal free-electron behaviour are directly accessible from knowledge of the dielectric permittivity [4.3], [4.4].

In principle, Maxwell theory can be applied in the whole frequency- and wave vector domain of electromagnetic waves, including the region of absorption edges, which we have dealt with in the foregoing chapters. Why, then, is Maxwell theory usually restricted to the low frequency region ($\hbar\omega \leq 50eV$)?

Since the high frequency energy loss spectrum ($\hbar\omega > 50\ eV$) shows excitation of strongly bound electrons, one could presume that Maxwell theory—which is a continuum theory—has to be replaced by quantum field theory in order to describe properly the quantum phenomena on the atomic scale of matter, and that absorption edge energy loss spectra are a paradigm of non-Maxwellian behaviour.

It should be observed, however, that Maxwell theory has nothing to do with a particular permittivity that is a bare phenomenological quantity. Apart from historical use, the very reason for the different treatment of loss spectra in the low and high energy region is that nothing is gained from a calculation of ε in the latter case. A direct comparison of the scattering cross section with theoretical predictions via Eq. (2.14) is straightforward.

Even in the low energy case not much is gained from a knowledge of ε compared with the scattering cross section, since there is a close correspondence between the scattering cross section and the dielectric function ε. We shall derive this relation at length in Chapter 5. At this point, it suffices to state that exactly the same information is contained in the cross section as is contained in the dielectric function. If there were standard models for the prediction of scattering cross sections in the low energy region, such as the hydrogenic model discussed in Chapter 2 for the high energy case, the concept of ε would be no longer of advantage in data interpretation. Unfortunately, the low energy response of a medium is much more intricate than the high energy response. Many-body effects play a dominant role, as we shall see, preventing a straightforward prediction of scattering cross sections. Experimenters are often content with a qualitative interpretation of their measurements in terms of the dielectric function [4.3], [4.4], [4.6], [4.7]. See Fig. 4.1 for the relations between models, experiment, cross section and ε.

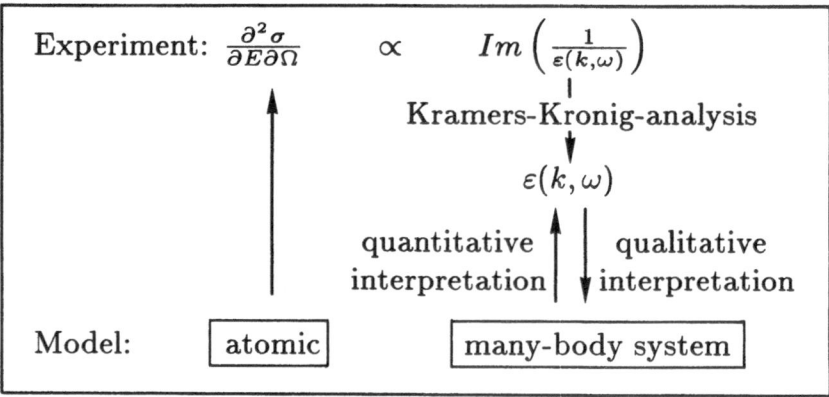

Fig. 4.1. The relations between scattering cross section, loss function and permittivity, and their connection to theory and experiment.

It is justified, therefore, to consider Maxwell theory a useful and elucidating link between the electromagnetic behaviour of a system and its dielectric function ε (as well as the permeability μ for magnetic materials). Further interpretation of ε may either proceed in a qualitative manner (position and linewidth of electronic transitions in the spectrum show up in a graph of $Im(\varepsilon)$ as exemplified in Fig. 4.2) or by comparison with theoretical predictions from quantum mechanical calculations (see Fig. 4.3).

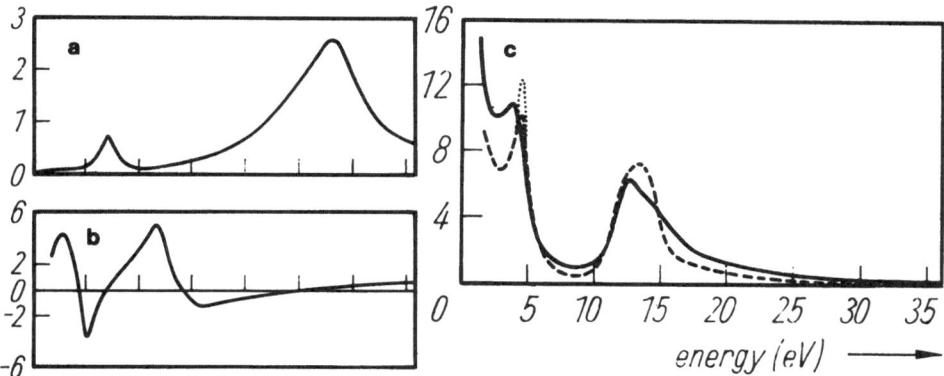

Fig. 4.2. a) Energy loss function $Im\left(-1/\varepsilon_\perp\right)$ of graphite derived from experimental inelastic scattering cross section;
b, c) dielectric function $\varepsilon_\perp = Re\left(\varepsilon_\perp\right) + \iota Im\left(\varepsilon_\perp\right)$ derived from a) by Kramers-Kronig analysis. Dashed and dotted lines in c) from optical measurements. The maxima in c) are caused by electronic interband transitions. From [4.3].

In the following, we shall investigate the most simple medium that provides insight into the electromagnetic response: the homogeneous, isotropic, nonmagnetic case. We shall review the Maxwell equations in a convenient form, namely in Fourier space, since eigenmodes and their features can be understood most easily this way. We shall not dwell on details, but rather present the familiar example of a metal with emphasis on the dispersion relation of the basic modes—longitudinal and transverse oscillations. The reader interested in more details of electromagnetic properties of conductive or non-conductive media is referred to textbooks on classical electrodynamics (e.g. [4.8]).

58

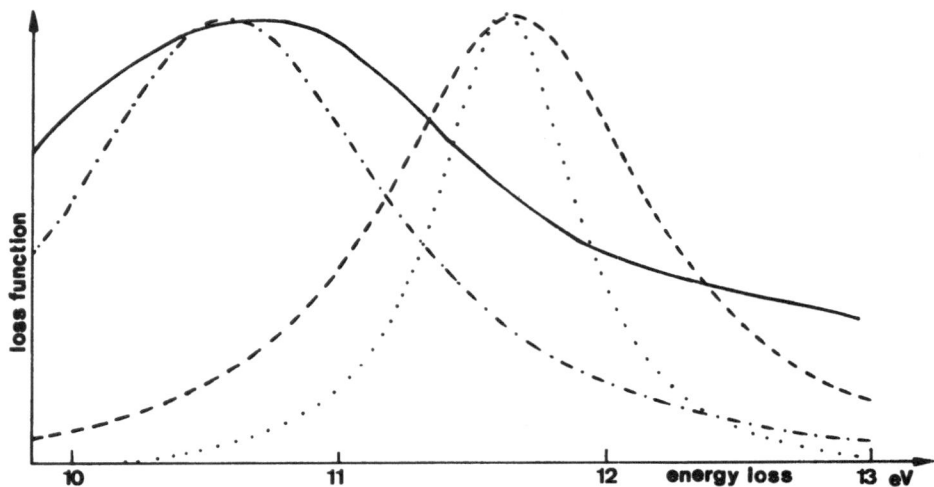

Fig. 4.3. Fittings of theoretical predictions of the loss function of amorphous MgZn to experimental results (bold line) by variation of model parameters. RPA-calculation with weak pseudopotential, ········
— Same—including a phenomenologic damping constant, ----
— Same—including the influence of Zn-d-band transitions. .—.—.—
From [4.5].

4.1. Fourier Transform

As a prerequisite, we recall some relations for Fourier transforms.

A superposition of plane waves can be given as

$$\tilde{f}(\vec{r},t) = \frac{1}{(2\pi)^4} \int\limits_{-\infty}^{+\infty} d\omega \int\limits_{-\infty}^{+\infty} d^3k\, f(\vec{k},\omega) e^{i(\vec{k}\vec{r}-\omega t)} \tag{4.1a}$$

where the denominator $(2\pi)^4$ has been introduced for convenience and f is the Fourier transform of $\tilde{f}$ times $(2\pi)^2$

$$f(\vec{k},\omega) = \int\limits_{-\infty}^{+\infty} dt \int\limits_{-\infty}^{+\infty} d^3r\, \tilde{f}(\vec{r},t) e^{-i(\vec{k}\vec{r}-\omega t)}. \tag{4.1b}$$

The relations between Fourier transforms, established by (4.1a,b) are sometimes expressed symbolically, as

$$\tilde{f}(\vec{r}, t) \leftrightarrow f(\vec{k}, \omega). \tag{4.2}$$

By taking the partial derivatives of Eq. (4.1a) with respect to t and the components of $\vec{r}$ it is at once evident that

$$\nabla \tilde{f} \leftrightarrow i k f, \tag{4.3a}$$

$$\frac{\partial}{\partial t} \tilde{f} \leftrightarrow -i\omega f \tag{4.3b}$$

which means that the operators ∇ or $\partial/\partial t$ in any expression are replaced by the respective factors when dwelling in Fourier space.

The convolution of two functions f, g is defined as

$$\tilde{f} * \tilde{g} := \tilde{h}(x) = \int_{-\infty}^{+\infty} \tilde{f}(x - x')\tilde{g}(x')dx'. \tag{4.4}$$

From (4.1a)

$$\begin{aligned}
h(k) &= \int \tilde{h}e^{-ikx}dx \\
&= \int dx' \int dx e^{-ikx} \tilde{f}(x - x')\tilde{g}(x') \\
&= \int dx' \tilde{g}(x') \int d\xi e^{-ik(\xi + x')} \tilde{f}(\xi) \\
&= \int dx' e^{-ikx'} \tilde{g}(x') \int d\xi e^{-ik\xi} \tilde{f}(\xi) \\
&= g(k) \cdot f(k)
\end{aligned} \tag{4.5}$$

Hence the convolution theorem

$$\tilde{f} * \tilde{g} \leftrightarrow f \cdot g \tag{4.6a}$$

holds, which is likewise valid for more than one variable. It is easy to show by the same argument that

$$\tilde{f} \cdot \tilde{g} \leftrightarrow \frac{1}{(2\pi)^n} f * g \tag{4.6b}$$

where n is the number of variables with respect to which the transform is carried out. Henceforth, our notation expresses r, t-dependence by a tilde ($\tilde{}$) over the function symbol.

4.2. Linear Response of a Medium

The response of a medium to an external electric field is usually defined through the polarization P (which is related to the polarization charge density, by div $P = \rho_{pol}$):

$$\tilde{\vec{P}} = \chi \tilde{\vec{E}} \tag{4.7}$$

where χ is the dielectric susceptibility. The response is described by χ or, what is equivalent, by the dielectric constant (permittivity, dielectric function)

$$\varepsilon = 1 + 4\pi\chi \tag{4.8}$$

and

$$\tilde{\vec{D}}(\vec{r}, t) = \tilde{\varepsilon}\tilde{\vec{E}}(\vec{r}, t). \tag{4.9}$$

Here, the properties of the medium are contained solely in the factor ε which couples fields $\vec{D}$ and $\vec{E}$.—We have already encountered a similar factorization in Chapter 2 in quite another context (see Eq. (2.23)).—The displacement field $\vec{D}$ may be viewed as a "driving agent" onto the medium which is characterized by ε. Imposing $\vec{D}$ results in a response of the medium, namely the polarization $\vec{P}$. It is important to realize at this point that the choice of a particular external field $\vec{D}$ will of course elicit a particular response, *but will by no means influence ε*.

Eq. (4.9) expresses that a) the response is linear in $\vec{E}$ and b) the polarization at $\vec{r}, t$ is neither influenced by the history of the system nor by the surroundings of $\vec{r}$—which is a fairly artificial assumption. In order to demonstrate this, consider the case that an external field is switched on immediately, as depicted in Fig. 4.4. Due to the inertia of the charge carriers in the medium, the polarization will gradually increase. Obviously, Eqs. (4.7) and (4.9) cannot be satisfied in this example for any $\varepsilon(t)$ whatsoever [4.9]. After a first glance at Fig. 4 it could be argued that $\varepsilon(t) = D(t)/E(t)$ for $t > t_0$. But, let the external field be switched on at a later instant: then, ε as a function of time will change—a fact which contradicts the assumption taken above that ε must not be influenced by the field D, esp. by the time at which D is activated. The same argument holds in $\vec{r}$-space.

Keeping the linearity between D and E, the simplest extension of Eq. (4.9) consistent with our example is

$$\tilde{\vec{D}}(\vec{r},t) = \int dt' \int d^3r' \, \varepsilon(\vec{r},\vec{r'},t,t')\tilde{\vec{E}}(\vec{r'},t').\qquad(4.10)$$

When the properties of the medium are time-independent, $\tilde{\varepsilon} = \tilde{\varepsilon}(t-t')$, and furthermore for isotropic, homogeneous systems we have

$$\tilde{\varepsilon} = \tilde{\varepsilon}(\vec{r}-\vec{r'},t-t').\qquad(4.11)$$

(Note that Eq. (4.9) is a particular case of Eq. (4.11) for a "local" and "immediate" response $\tilde{\varepsilon} = \varepsilon_i \cdot \delta^3(\vec{r}-\vec{r'})\delta(t-t')$ The subscript denotes "immediate").

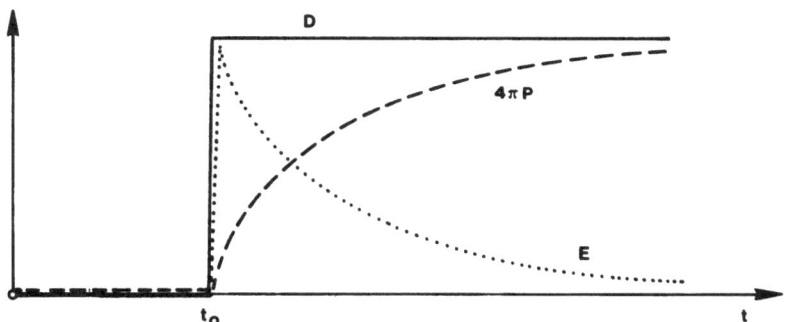

Fig. 4.4. Delayed response $4\pi P$ (dashed line) of a medium to an external perturbation D (solid line). The total field is E (dotted line). By changing the instant of activating D, it can be shown that $\varepsilon = D/E$ depends on the perturbation, contrary to the assumption taken (see the text above).

Eq. (4.10) with (4.11) is recognized as a convolution in space and time, hence

$$\tilde{\vec{D}} = \tilde{\varepsilon} * \tilde{\vec{E}} \leftrightarrow \vec{D}(\vec{k},\omega) = \varepsilon(\vec{k},\omega)\vec{E}(\vec{k},\omega).\qquad(4.12a)$$

(Note again, that the local and immediate response (4.9) is represented as $\vec{D}(\vec{k},\omega) = \varepsilon_i \vec{E}(\vec{k},\omega)$ in Fourier space). Besides ε, the behaviour of the system is described by the magnetic permeability and the conductivity σ, for which, in linear response

$$\tilde{\vec{B}} = \tilde{\mu} * \tilde{\vec{H}} \leftrightarrow \vec{B} = \mu \cdot \vec{H}\qquad(4.12b)$$

$$\tilde{\vec{j}} = \tilde{\sigma} * \tilde{\vec{E}} \leftrightarrow \vec{j} = \sigma \cdot \vec{E} \qquad (4.12c)$$

hold.

For the sake of clarity, we mention an almost trivial result: The vector property remains unchanged by Fourier transforms. As an example, assume

$$\tilde{\vec{E}}(\vec{r},t) = \vec{E}_0 \cdot e^{i(\vec{k}_0 \vec{r} - \omega t)} \leftrightarrow \vec{E}(\vec{k},\omega) =$$

$$= \vec{E}_0 \iint dt \, d^3 r \, e^{i(\vec{k}_0 - \vec{k})\vec{r}} e^{-i(\omega - \omega_0)t} =$$

$$= \vec{E}_0 (2\pi)^4 \cdot \delta^3(\vec{k} - \vec{k}_0) \cdot \delta(\omega - \omega_0) \qquad (4.13)$$

which is to say, the direction of a vector field with harmonic variation equals that of its Fourier transform.—More details on linear response can be found in [2.4].

4.3. The Maxwell Equations

Now we are prepared to formulate the Maxwell equations in Fourier representation:

$$\nabla \tilde{\vec{D}} = 4\pi \tilde{\rho}_{ex} \qquad \leftrightarrow \qquad i(\vec{k}, \vec{D}) = 4\pi \rho_{ex} \qquad (4.14a)$$

$$\nabla \tilde{\vec{B}} = 0 \qquad \leftrightarrow \qquad i(\vec{k}, \vec{B}) = 0 \qquad (4.14b)$$

$$\nabla \times \tilde{\vec{E}} = -\frac{1}{c}\frac{\partial}{\partial t}\tilde{\vec{B}} \qquad \leftrightarrow \qquad i(\vec{k} \times \vec{E}) = \frac{i\omega}{c}\vec{B} \qquad (4.14c)$$

$$\nabla \times \tilde{\vec{H}} = \frac{4\pi}{c}\tilde{\vec{j}}_{ex} + \frac{1}{c}\frac{\partial}{\partial t}\tilde{\vec{D}} \leftrightarrow i(\vec{k} \times \vec{H}) = \frac{4\pi}{c}\vec{j}_{ex} - \frac{i\omega}{c}\vec{D} \qquad (4.14d)$$

We prefer to use the electrostatic system of units ($\varepsilon_0 = \mu_0 = 1$) instead of the SI-system because the displacement field $\vec{D}$ is of the same dimension as $\vec{E}$ in the former. This will be of some advantage in the discussion of the possible eigenmodes of a system, as we shall see shortly.

Common use has wiped out much of the feeling for the vividness and beauty inherent in the Maxwell equations. In 1893, L. Boltzmann commented on Eqs. (4.14) when he said: "War es ein Gott, der diese Zeilen schrieb ..." [4.16] ("Was it a god who wrote these lines...". Ever since the aestetic attitude of the physicist has gradually faded away in favour of a sound understanding.

To work out the advantages of the Fourier representation we start with a repetition of how (4.14) is usually solved in the source-free case when the external current density vanishes, $\vec{j}_{ex} = \vec{0}$, as does the external charge density: $\rho_{ex} = 0$. From (4.14b,a), $\vec{B}$ and $\vec{E}$ are given in terms of vector and scalar potentials $\vec{A}, \varphi$ as

$$\tilde{\vec{B}} = \nabla \times \tilde{\vec{A}} \leftrightarrow \vec{B} = i\vec{k} \times \vec{A}, \qquad (4.15a)$$

$$\tilde{\vec{E}} = -\frac{1}{c}\frac{\partial}{\partial t}\tilde{\vec{A}} - \nabla \tilde{\varphi} \leftrightarrow \vec{E} = \frac{i\omega}{c}\vec{A} - i\vec{k}\varphi. \qquad (4.15b)$$

Additionally, we gauge the field by the Coulomb or transverse gauge

$$\nabla \tilde{\vec{A}} = 0 \leftrightarrow i\vec{k}\vec{A} = 0. \qquad (4.15c)$$

Now, insert Eqs. (4.15a,b) in (4.14d,a) assuming local and immediate response (4.9) with $\varepsilon(k,\omega) = \varepsilon_i$, $\mu(k,\omega) = \mu_i$:

$$\nabla \times (\nabla \times \tilde{\vec{A}}) = \frac{\varepsilon_i \mu_i}{c}\frac{\partial}{\partial t}(-\frac{1}{c}\frac{\partial \tilde{\vec{A}}}{\partial t} - \nabla \tilde{\varphi}) \leftrightarrow$$

$$\leftrightarrow i\vec{k} \times (i\vec{k} \times \vec{A}) = \frac{-i\omega}{c}\varepsilon_i \mu_i(\frac{i\omega}{c}\vec{A} - i\vec{k}\varphi). \qquad (4.16a)$$

$$\nabla[\varepsilon_i(-\frac{1}{c}\frac{\partial \tilde{\vec{A}}}{\partial t} - \nabla \tilde{\varphi})] = 0 \leftrightarrow i\vec{k}\varepsilon_i(\frac{i\omega}{c}\vec{A} - i\vec{k}\varphi) = 0. \qquad (4.16b)$$

Observe $\vec{a} \times (\vec{b} \times \vec{c}) = \vec{b}(\vec{a}\vec{c}) - \vec{c}(\vec{a}\vec{b})$ and (4.15c), then from (4.15a,b)

$$\nabla^2 \tilde{\vec{A}} - \frac{\varepsilon_i \mu_i}{c^2}\frac{\partial^2}{\partial t^2}\tilde{\vec{A}} - \frac{\varepsilon_i}{c}\frac{\partial}{\partial t}\nabla \tilde{\varphi} = \vec{0} \leftrightarrow \qquad (4.17a)$$

$$\leftrightarrow k^2\vec{A} - \frac{\varepsilon_i \mu_i \omega^2}{c^2}\vec{A} + \frac{\varepsilon_i \omega \vec{k}}{c}\varphi = \vec{0},$$

$$\varepsilon_i \nabla^2 \tilde{\varphi} = 0 \leftrightarrow \varepsilon_i \vec{k}^2 \varphi = 0. \qquad (4.17b)$$

From (4.17b)

$$\tilde{\varphi} = const \leftrightarrow \varphi = const \cdot \delta^3(\vec{k}), \qquad (4.18)$$

and (4.17a) yields the wave equation

$$\Delta \vec{\tilde{A}} - \frac{\varepsilon_i \mu_i}{c^2} \frac{\partial^2}{\partial t^2} \vec{\tilde{A}} = \vec{0} \leftrightarrow (k^2 - \frac{\varepsilon_i \mu_i \omega^2}{c^2})\vec{A} = \vec{0} \qquad (4.19)$$

the solutions of which are

$$\vec{\tilde{A}} = \int d\omega \int d^3k \vec{A}(\vec{k},\omega) e^{i(\vec{k}\vec{r}-\omega t)} \leftrightarrow \vec{A}(k,\omega) \qquad (4.20)$$

with arbitrary $\vec{A}$. From the Fourier representation (4.19), the frequency dispersion law

$$\frac{\varepsilon_i \mu_i}{c^2}\omega^2 = k^2 \qquad (4.21)$$

is at once evident for $\vec{A}$ supposed to be nontrivial. From (4.15b) it follows that $\vec{E}\|\vec{A}$, and since $\vec{k}\vec{A} = 0$—(see Eq. 4.15c)—, we conclude $\vec{k}\perp\vec{A}$, hence $\vec{E}\perp\vec{k}$, the electric field is normal to the wave vector, and from (4.15a) $\vec{B}\perp\vec{E}$, $\vec{B}\perp\vec{k}$. So, we have derived the commonplace that electromagnetic waves *must* be transverse. Seemingly there are no longitudinal solutions to the Maxwell equations.

Note that this was due to Eq. (4.17b)

$$\varepsilon_i \nabla^2 \tilde{\varphi} = 0 \leftrightarrow \varepsilon_i \vec{k}^2 \varphi = 0, \qquad (4.17b)$$

so, $\nabla\tilde{\varphi}$ vanished. Virtually, (4.17b) tells that either $\tilde{\varepsilon}_i \neq 0$ and $\nabla\tilde{\varphi} = 0$ or $\varepsilon_i = 0$ and $\nabla\tilde{\varphi}$ is arbitrary! Of course, for local, immediate response it does not make sense to put $\varepsilon_i = 0$ since there is no such medium. However, for the derivation in Fourier space from (4.14) to (4.21), local immediate response need not be assumed. All formulas on the right hand side apply as well when ε, μ are nonlocal in space and time: $\varepsilon = \varepsilon(\vec{k},\omega)$, $\mu = \mu(\vec{k},\omega)$. And there might be cases where ε vanishes for particular (k_0,ω_0). For such exotic cases, Eq. (4.17b) tells that $\varphi(k_0,\omega_0)$ is arbitrary and from (4.17a) we have $\vec{A}(k_0,\omega_0) = 0$. In that case we infer from Eq. (4.15b) that $\vec{E}\|\vec{k}$. So there *could* be longitudinal electric waves. But what about the magnetic field? Are all equations consistent with $\vec{E}\|\vec{k}$? We anticipate the result that longitudinal charge density waves exist as a rule, not as an exotic exception. Most interestingly, these modes have only been treated in more recent textbooks on solid state physics e.g. [4.11], although they were investigated as early as 1926 [4.1], [4.23] in gaseous plasmas. The famous textbook of Arnold Sommerfeld [4.12] on theoretical physics or the standard monograph of A. Sommerfeld and H. Bethe [2.2] on electron theory of metals do not even mention longitudinal waves. It was not ignorance, but the lack of experimental possibilities and the alleged unimportance of longitudinal electric waves which led to their being overlooked, until 1941 Ruthemann observed plasmons in energy loss spectra of thin metal foils for the first time—without exactly knowing what he observed [4.14], [4.15].

In the sequel, we shall disarray the problems just addressed by reference to the coordinate system sketched in Fig. 4.5.

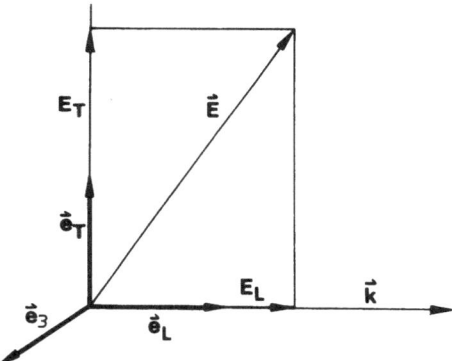

Fig. 4.5. Reference frame for the vector representation (4.22) of the Maxwell Equations.

As can be shown in a few lines, the Maxwell Equs. (4.14c,d) in Fourier representation read, with respect to the frame Fig. 4.5 and in matrix algebra:

$$\begin{pmatrix} \frac{\omega\varepsilon}{c} & 0 & 0 & 0 \\ 0 & \frac{\omega\mu}{c} & 0 & 0 \\ 0 & 0 & \frac{\omega\varepsilon}{c} & -k \\ 0 & 0 & -k & \frac{\omega\mu}{c} \end{pmatrix} \begin{pmatrix} E_l \\ H_l \\ E_t \\ H_3 \end{pmatrix} = \begin{pmatrix} \frac{4\pi i}{c} j_l \\ 0 \\ -\frac{4\pi i}{c} j_t \\ 0 \end{pmatrix}. \qquad (4.22)$$

(Eqs. (4.14a,b) follow from (4.14c,d) by application of ∇ and the continuity equation).

In the source-free case the right-hand side of Eq. (4.22) vanishes. Denoting the column vector by $\vec{X}$, and the matrix by $\mathbf{C}$, (4.22) is

$$\mathbf{C}\vec{X} = \vec{0}. \tag{4.23}$$

Solutions exist when

$$Det\ \mathbf{C} = \omega^2 \varepsilon \mu [\frac{\omega^2 \varepsilon \mu}{c^2} - k^2] = 0. \tag{4.24}$$

There are four othogonal eigenvectors $\vec{X}_j$, each defined by a root of the characteristic Eq. (4.24), namely

$$root : \varepsilon(k,\omega) = 0 \quad \mu(k,\omega) = 0 \quad k = \frac{\omega}{c}\sqrt{\varepsilon\mu} \quad k = -\frac{\omega}{c}\sqrt{\varepsilon\mu}$$

$$\begin{pmatrix} E_l \\ 0 \\ 0 \\ 0 \end{pmatrix} \quad \begin{pmatrix} 0 \\ H_l \\ 0 \\ 0 \end{pmatrix} \quad \begin{pmatrix} 0 \\ 0 \\ E_t^+ \\ H_3^+ \end{pmatrix} \quad \begin{pmatrix} 0 \\ 0 \\ E_t^- \\ H_3^- \end{pmatrix}. \tag{4.25}$$

Each of the four fundamental waves can exist at its "frequency dispersion law" $\omega = \omega(k)$ given implicitly by the roots of (4.24).

A similar coordinate-independent formulation can be found, for instance, in [5.6].

There are pure longitudinal electric and magnetic waves besides the well-known transverse electromagnetic ones. In the following we shall assume nonmagnetic systems, i.e. $\mu(k,\omega) \equiv 1$, hence the magnetic longitudinal solution is discarded here.—An important difference between longitudinal (electric) and transverse waves is seen from application of the continuity equation onto the polarization current $\vec{j}_{pol}$ which is parallel to $\vec{E}$: Since the change in polarization charge density is proportional to the divergence of $\vec{j}_{pol}$, the charge density remains constant for transverse waves. Longitudinal waves demand periodic charge density fluctuations in the medium. Since they were predicted and detected for the first time in plasmas by Langmuir [4.1], [4.2] they have been coined *plasmons* in the late fifties in analogy to phonons, the quantized lattice oscillations.

4.4. The Dielectric Function

Before considering ε, a remark should be added on the relation between ε and σ. $\vec{j}_{ex}$ in (4.14d) are external (non-polarization) currents. We can also write

$$i(\vec{k} \times \vec{H}) = \frac{4\pi}{c}(\vec{j}_{ex} + \vec{j}_{pol}) - \frac{i\omega}{c}\vec{E} = \frac{4\pi}{c}\vec{j}_{ex} - \frac{i\omega\varepsilon\vec{E}}{c}. \qquad (4.14e)$$

Ohm's law relates $\vec{j}_{pol}$ and $\vec{E}$, as $\vec{j}_{pol} = \sigma\vec{E}$. Hence, insertionn of $\vec{j}_{pol}$ and comparison of the coefficients of $\vec{E}$ in Eq. (4.14e) yields

$$\varepsilon = 1 + \frac{4\pi i\sigma}{\omega} \qquad (4.26)$$

which is to say: for nonmagnetic materials, ε describes the system completely. For metals $\sigma(\vec{k},0) \neq 0$, and, $\varepsilon(\vec{k},0) \to \infty$, in which case the *d.c.* conductivity is necessary to describe the response of the medium.

In any case, apart from $\omega = 0$, the dielectric function $\varepsilon(\vec{k},\omega)$ totally describes the response of the system to electromagnetic disturbances. From optics, we know there is an intimate relation between ε and the complex index of refraction $n + i \cdot \kappa$

$$n^2 - \kappa^2 = Re(\varepsilon), \quad 2n\kappa = Im(\varepsilon). \qquad (4.27)$$

So ε determines the electrical and optical properties of a medium. (On the other hand, $\varepsilon(\vec{k},\omega)$ *is* determined by electronic properties, such as interband transitions, absorption edges, resonant dipole excitation).

4.5. The Drude Model

Next we consider a simple model in order to conceive of important characteristics of the permittivity ε. We assume an ideal metal without any resonant frequencies, i.e., the conduction electrons behave as if they were free.

The equation of motion for the conduction electrons of mass m in a longitudinal field $\vec{E}$ is

$$m\frac{\partial \vec{\tilde{v}}}{\partial t} + m\frac{\vec{\tilde{v}}}{\tau} = e\vec{\tilde{E}} \leftrightarrow -i\omega m\vec{v} + \frac{m}{\tau}\vec{v} = e\vec{E} \qquad (4.28)$$

where e is the charge of an electron, $\vec{v}$ its velocity and τ a characteristic relaxation time due to friction in the electron gas. Note that

$$\vec{j} = \rho_0 \vec{v} = ne\vec{v}. \qquad (4.29)$$

ρ_0 is the equilibrium charge density and n the equilibrium number density of electrons.

Next, multiply Eq. (4.28) by $\rho_0/m = ne/m$,

$$\frac{\partial \vec{\tilde{j}}}{\partial t} + \frac{\vec{\tilde{j}}}{\tau} = \frac{ne^2}{m}\vec{\tilde{E}} \leftrightarrow \vec{j}(\frac{1}{\tau} - i\omega) = \frac{ne^2}{m}\vec{E}. \qquad (4.30)$$

Here, $\vec{j}_{ex} = \vec{0}$ and $\vec{j} = \vec{j}_p$ is a polarization current (i.e. the response of the system) when $\vec{E}$ is caused by an external perturbation. $\vec{j}$ is related to the dipole moment density $\vec{\tilde{P}} = ne(\vec{r}(t) - \vec{r}_0)$, where $\vec{r}_0$ is the equilibrium position of an electron, by (4.29) as

$$\vec{\tilde{j}} = ne\frac{\partial \vec{r}(t)}{\partial t} = \frac{\partial}{\partial t}\vec{\tilde{P}}(t) \leftrightarrow \vec{j} = -i\omega\vec{P}. \qquad (4.31)$$

Now, from (4.8), (4.9) and from (4.30), (4.31)

$$\varepsilon = 1 + \frac{4\pi|\vec{P}|}{|\vec{E}|} = 1 + \frac{4\pi ne^2}{m}\frac{1}{-i\omega(\frac{1}{\tau} - i\omega)}. \qquad (4.32)$$

Observe that ε is independent of $\vec{k}$ since the constituting Eq. (4.28) did not contain any space-dependent part. Defining the plasma frequency

$$\omega_p^2 = \frac{4\pi ne^2}{m}, \qquad (4.33)$$

ε, separated into $Re(\varepsilon)$ and $Im(\varepsilon)$, is

$$\varepsilon_1 := Re(\varepsilon) = 1 - \frac{\omega_p^2}{\omega^2 + \frac{1}{\tau^2}} \qquad (4.34a)$$

$$\varepsilon_2 := Im(\varepsilon) = \frac{1}{\omega\tau} \frac{\omega_p^2}{\omega^2 + \frac{1}{\tau^2}} \qquad (4.34b)$$

These "Drude expressions" for the dielectric function of a metal are displayed in Fig. 4.6. As can be seen, the inclusion of a friction term $1/\tau$ shifts the zero of ε_1 from ω_p to $\sqrt{\omega_p^2 - 1/\tau^2}$. By the same token, an imaginary part of ε occurs (which was to be expected since ε_2 always describes dissipation of energy in a system, here caused by the friction term $\vec{j}/\tau$.)

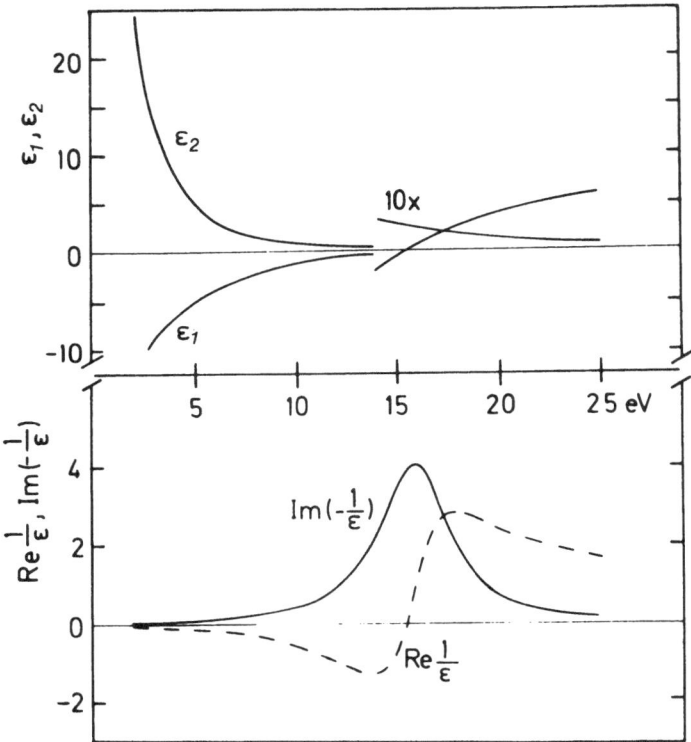

Fig. 4.6. Dielectric function and loss function $Im(-1/\varepsilon)$ of a free electron gas for $\hbar\omega_p = 15\ eV$ and $\hbar/\tau = 4\ eV$. From [4.10].

It should be mentioned that in the derivation given above it was assumed that the system is isotropic (otherwise $\chi = P/E$ would depend on the direction of $\vec{E}$, so the susceptibility χ would be a tensor) and homogeneous (otherwise local field effects would occur, see Chapt. 8). Deviation of n, κ from the Drude behaviour (4.34) is significant for the existence of resonant oscillations [4.5]. In this case, optical observations cannot be described by (4.27) with (4.34). (For instance, when the medium exhibits one undamped resonant oscillator at ω_t, ε is given by

$$\varepsilon = 1 + \frac{f}{\omega_t^2 - \omega^2} \tag{4.35}$$

instead of Eq. (4.34). (f is the oscillator strength). Obviously, when $\omega_t \to 0$, the Drude behaviour for $\tau \to \infty$ is obtained (To put it another way: conduction electrons in a metal have resonant frequency $\omega_t = 0$.

The relaxation time τ relates to the conductivity σ as follows: For stationary (d.c.) conditions, $\partial \vec{j}/\partial t = 0$. Hence, the differential equation (4.30) for the current is

$$\frac{\vec{j}}{\tau(0)} = \frac{ne^2}{m}\vec{E}. \tag{4.36}$$

Ohm's law yields

$$\tau(0) = \sigma_{dc}\frac{m}{ne^2} = \frac{4\pi\sigma_{dc}}{\omega_p^2} \tag{4.37}$$

(in fact, τ is twice this value. This is due to the fact that even in equilibrium, electrons are accelerated until they suffer collisions, so microscopically, $\partial \vec{j}/\partial t \neq 0$).

4.6. Charge Oscillations in a Metal

Now, the question arises as to what the solutions of the Maxwell Equs. (4.14) look like in a homogeneous, infinite metal described by the Drude expressions (4.34). In a first approximation, we assume $\tau \to \infty$, i.e. $\varepsilon_2(\omega) = 0$. Then, for the electromagnetic transverse modes, two regions can be discerned: for $\varepsilon_1 > 0$ the wave number k is a real function of real ω, whereas for $\varepsilon_1 < 0$ it is purely imaginary, due to (4.21):

$$k = \frac{\omega}{c}\sqrt{\varepsilon_1 \mu} = \frac{\omega}{c}\sqrt{\varepsilon_1} \qquad \begin{matrix} \epsilon R & \varepsilon_1 > 0 \\ imag. & \varepsilon_1 < 0 \end{matrix} \qquad (4.38)$$

The real branch describes propagating, undamped waves. In Fig. 4.7, a (k, ω)-diagram is drawn.

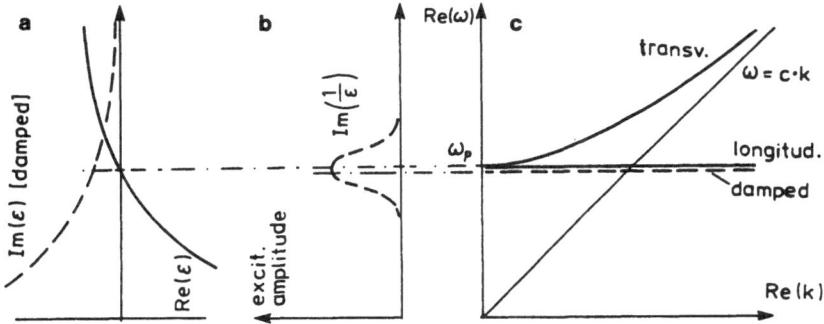

Fig. 4.7. a) Drude expressions for $\varepsilon(\omega)$, in a metal,
b) corresponding loss function peaking at the zero of $Re(\varepsilon)$,
c) corresponding dispersion of longitudinal and transverse modes. Damping shifts the longitudinal mode to lower frequencies.

For $\varepsilon_1 < 0$, i.e. $\omega < \omega_p$, the transverse modes are damped and do not propagate (purely imaginary k), but the metal is transparent for $\omega > \omega_p$. At ω_p, longitudinal electric waves exist for any k. This is the frequency dispersion for the collective oscillation. Note that the phase velocity of the longitudinal mode, $c_{ph} = \omega/k$ decreases from ∞ to 0 as k increases, and its group velocity $\partial\omega/\partial k$ vanishes. This result implies that the longitudinal mode cannot transport any energy.—The transverse group velocity increases from 0 to c.

72

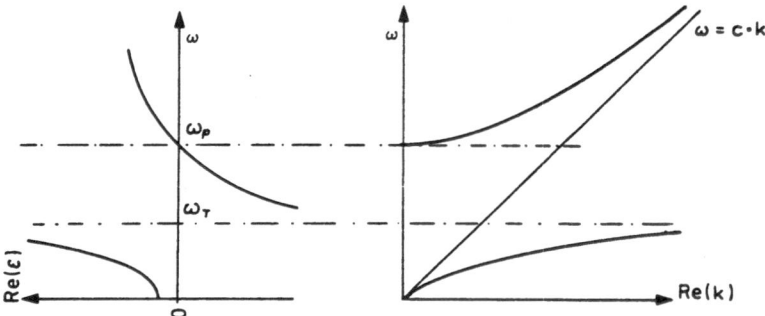

Fig. 4.8. Dispersion of the transverse fields in the presence of a transverse oscillation at $\omega = \omega_t$ $(\omega_t > 0)$. The longitudinal mode is $\omega(k) = \omega_p$.

Since $\mu(k, \omega) = 1$, there is no magnetic longitudinal oscillation.

When $\varepsilon_2 \neq 0$, (4.21) tells that k^2 has an imaginary part; hence k is complex even for $\omega > \omega_p$, i.e. transverse waves are always damped in a metal with finite conductivity (see (4.37)).

The longitudinal oscillation E_l for finite τ can be derived from setting zero ε in Eq. (4.32).

$$\varepsilon = 1 - \frac{\omega_p^2}{i\omega(\frac{1}{\tau} - i\omega)} = 0. \tag{4.39}$$

It is satisfied by

$$\omega = \sqrt{\omega_p^2 - (\frac{1}{2\tau})^2} - \frac{i}{2\tau} \tag{4.40}$$

which is a decaying oscillation. Since $E_l \propto e^{-i\omega t}$, the lifetime of the excited mode is 2τ. Plotted as the excitation amplitude-vs-frequency, the spectral distribution of the longitudinal mode peaks at $Re(\omega) = 0$ and has a width of $1/2\tau$ (see Fig. 4.7).

We mention that in case of an oscillator at ω_t, the real part ε_1 is given by (4.35). Such a medium is transparent for $\omega\epsilon[0, \omega_t]$ and $\omega\epsilon[\omega_p, \infty]$, for transverse electromagnetic radiation (see Fig. 4.8). The group and phase velocities below ω_t decrease from c to 0.

From Fig. 4.8, we conclude that the condition $\varepsilon = 0$ can still be fulfilled when there are bound electrons in the medium, since any oscillator in the spectrum can be interpreted as a bound charge. So charge density waves (plasmons, longitudinal electric fields) even occur in non-metallic media.

When the plasmon is excited, the bound charges become polarized and oscillate at the longitudinal eigenfrequency. In metals and semiconductors, $\hbar\omega_p$ is of order (1-10) eV.

Instead of electrons, ions may oscillate longitudinally. The results derived here hold as well, with ω_p being replaced by the longitudinal optical phonon (LO) frequency. $\hbar\omega_{LO}$ is of order (0.01-0.1) eV, due to the much higher inertia of ions.

It is worthwhile noting that plasmon and photon modes interact with one another. Inspection of Fig. 4.8 shows that the light line $\omega = ck$ (the free phonon "dispersion") is strongly altered at the intersection with the plasmon line. The effect is in complete analogy with the hybridisation of the "free" photon with phonons. The hybrid mode is called polariton, more exactly plasmon polariton in order to discern it from phonon or other hybrids. (Fig. 4.9 is an example for the dispersion of a phonon polariton).

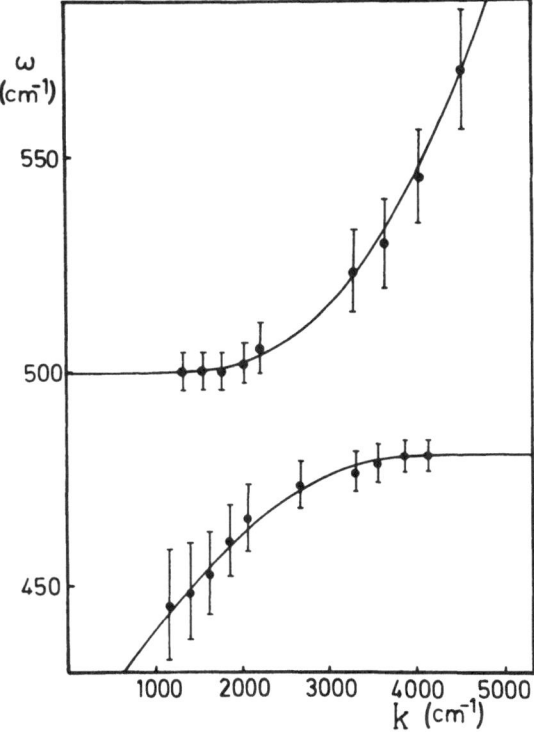

Fig. 4.9. Dispersion of the phonon polariton in fused quartz. From [4.17].

The polariton emerges from a crossover of two dispersion lines, i.e. *two* excitations are effective at the same frequency and momentum. The state of the system is degenerate at this particular frequency. As is well known from the physics of oscillations, degenerate states split when there is interaction between the states—such as with a coupled pendulum, Bragg-reflection or atomic orbital overlap. The number of branches in the dispersion graph remains constant when interaction is activated. In fact, Figs. 4.8, 4.9 reveal that there are two polariton branches: the lower one is a hybrid with the transverse oscillator at ω_t (it may be a TO-phonon or an electronic interband transition), the upper one is in a sense also caused by the same oscillator since there is no LO-mode without the corresponding TO-mode. In a free electron gas, $\omega_t = 0$, hence only the upper branch remains.

In closing this chapter, we mention that exciton polaritons play a central role in recent theories of luminescence. The reader interested in details is referred to the monograph of Agranovich and Galanin [4.13].

5. Some Details on Charge Oscillations

In the present chapter, we shall discuss longitudinal oscillations of charges in a medium more closely, starting from an intiutive picture of what happens when a perturbation is activated in the medium. Due to the Coulomb interaction between charges, there will be a tendency to screen off any disturbance. This screening effect is intimately connected with the occurence of charge density waves. In the sequel we shortly discuss the influence of boundaries and eventually the response of the system to an external perturbation. The issue of a fast probe electron entering the medium is calculated in detail. We shall derive an important relation between the differential scattering cross section and the dielectric permittivity ε.

5.1. Screening

The longitudinal charge density waves derived previously are closely related to the phenomenon of electrostatic screening. In order to elucidate this relation we assume that in a macroscopically homogeneous gas of charges a pointlike disturbance $\tilde{\rho}_{ex}(\vec{r}) = Q\delta^3(\vec{r})$ is inserted. As soon as $\tilde{\rho}_{ex}$ has been activated, the neighbouring charges will rearrange in a tendency to screen off the disturbance (Fig. 5.1).

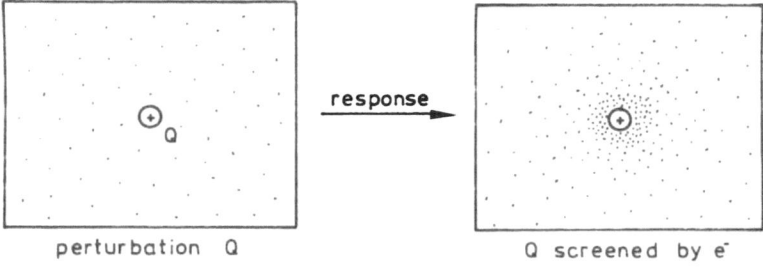

perturbation Q   Q screened by e⁻

Fig. 5.1. Static screening cloud of an external perturbation.

As a consequence, a charge density $\tilde{\rho}_{in}$ will be induced, and the screened potential is given by the Poisson equation

$$-\nabla^2 \tilde{\varphi} = 4\pi(\tilde{\rho}_{ex} + \tilde{\rho}_{in}) \longleftrightarrow k^2\varphi = 4\pi(Q + \rho_{in}) \qquad (5.1)$$

On the other hand, from thermodynamics we note that the total particle density $\tilde{n}$ in a potential $\tilde{\varphi}$ deviates from the equilibrium density $\tilde{n}_0$ by the Boltzmann-factor [2.4], [4.8]

$$\tilde{n} = \tilde{n}_0 e^{-e\tilde{\varphi}/\kappa T}. \qquad (5.2a)$$

Muliplying by e and substracting $\tilde{\rho}_0$ yields

$$\tilde{\rho}_{in} = \tilde{\rho}_0(e^{-e\tilde{\varphi}/\kappa T} - 1) \qquad (5.2b)$$

which can be approximated for small disturbances by an expansion to first order

$$\tilde{\rho}_{in} \simeq -\tilde{\rho}_0 e\tilde{\varphi}/\kappa T \longleftrightarrow \rho_{in} = -n_0 e^2 \varphi/\kappa T. \qquad (5.3)$$

n_0 is the equilibrium number density of the gas. Eqs. (5.1) and (5.3) yield, in k-representation

$$\varphi = \frac{4\pi Q}{k^2 + 4\pi n_0 e^2/\kappa T}. \qquad (5.4)$$

Note that φ lacks the singularity of the original Coulomb potential $4\pi Q/k^2$ at $k = 0$, i.e. the long-wavelength ($\equiv$ longrange) contributions of the Coulomb potential are screened off. Fig. 5.2 shows that $\varphi(r)$ is of the Yukawa form $\propto e^{-r/\lambda_s}$. The screening length λ_s, within which the Coulomb potential of a perturbation is attenuated to e^{-1} of its original value, is given by

$$\lambda_S = \sqrt{\kappa T/4\pi n_0 e^2} = \frac{\bar{v}}{\omega_p\sqrt{3}} \qquad (5.5)$$

as can be seen from Eq. (5.4). The last equality stems from replacement of $3\kappa T/2$ by $m\bar{v}^2/2$. This expression for λ_S was first derived by Debye and Hückel in 1923 [5.1] and is called Debye-length.

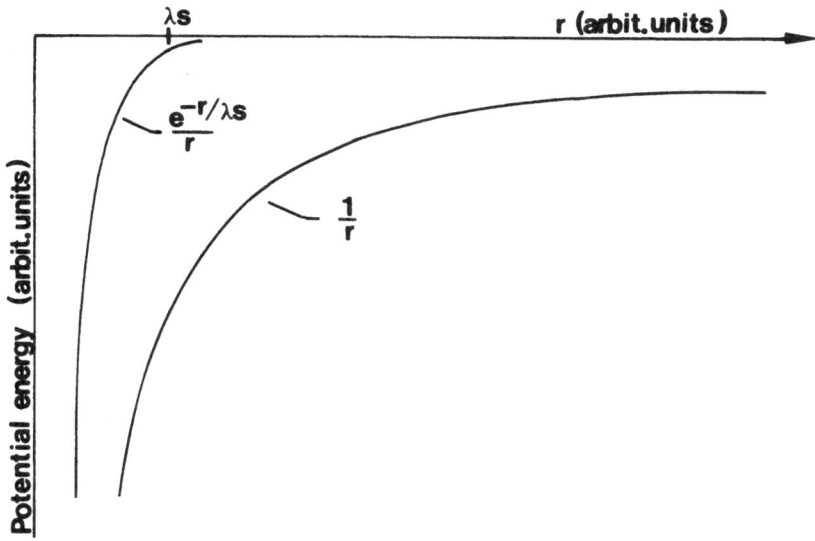

Fig. 5.2. Screening of the Coulomb potential of a pointlike perturbation. The screened potential is confined to a distance of $\sim \lambda_S$.

We note that in degenerate plasmas Eq. (5.5) no longer holds; instead, the screening distance is given by the Thomas-Fermi length (obtained by replacing $3\kappa T/2$ with E_F)

$$\lambda_{TF} = \sqrt{E_F/6\pi n e^2} = \frac{v_F}{\omega_p \sqrt{3}}. \tag{5.6}$$

As an example, we give an estimate from Eq. (5.6) of the screening length in metals: approximately, $v_F \approx 10^6 \ m/s$ and $\omega_p \approx 5 \cdot 10^{15} \ Hz$. Hence, $\lambda_{TF} \sim 1$ Å. Any external perturbation (charge or potential) is screened off within atomic distances. This is a more precise formulation of the famous result that in metals there are no macroscopic electrostatic fields.

We also note that the screened potential in a degenerate plasma has a long-range oscillatory part known as Friedel-Kohn wiggle or Friedel oscillation [5.2], [5.3]. The effect is caused by the Pauli exclusion principle (occurrence of a sharp Fermi surface) and is very sensitive to the shape of the surface [5.4].

5.2. Plasmons

It is easy to realize what happens when a perturbation is suddenly "switched on": Tending to screen off the perturbation, the charge carriers will overshoot their new equilibrium position—the screening cloud starts an oscillatory movement. When there is no damping, this movement resembles an eigenmode of the system. Since this "breathing" motion of the screening cloud is radially symmetric, there is no dipole or any multipole moment, and the magnetic field vanishes. According to Eq. (4.14c), $\vec{k} \times \vec{E} = \vec{0}$, hence $\vec{k} \| \vec{E}$, the screening oscillation is a longitudinal one. Previously, we showed that there is one longitudinal eigenmode resonant at a frequency $\omega = \omega_p$, called plasmon. Now it becomes evident that the plasmon is identical with the "breathing" motion of the screening cloud, caused by the eagerness of the charges in the metal to screen any potential. In fact, screening occurs in all types of plasma. The better the charges can move, the stronger is the screening effect, and the better defined is the plasmon. There is a direct relation between screening capabilities and occurrence of longitudinal oscillations in a medium. This fact indicates a convenient view of the electron gas as a many-body system: Instead of electrons with Coulomb interaction, one may also think of the system as consisting of quasi-particles exerting a screened (Yukawa-) potential onto one another. Additionally, in the new aspect, there is a collective motion (the plasmon).—We shall have opportunity to discuss quasi-particles in Chapter 7.

By use of quasi-particles and a collective mode the long-range Coulomb force has been done away with. The full advantage of this line of thought becomes obvious in the quantum mechanical treatment. As we shall see, the celebrated perturbation theory fails in the case of the electron gas. The failure stopped theoretical development for more than one decade. The idea of introducing quasi-particles and collective modes—emerging only in the fifties [7.5] eventually proved successful in overcoming the divergence problems.

As we have seen, the plasmon is an oscillation defined coherently throughout the medium. So it is a collective mode where all charge carriers participate. For the collective of charges to oscillate it is necessary that they exert forces onto one another. Since forces fall off rapidly within a sphere of radius λ_S around a charge carrier, this means that whenever there is more than one particle within approx. λ_s^3, or $n\lambda_s^3 > 1$, collective effects such as plasmons can be expected to occur *without external impetus*. From Eqs. (5.5), (5.6) the densities and temperatures at which plasmons will occur, can be calculated for both classical and degenerate plasmas [5.7], [5.11]. The results are subsummed in the log-log plot fig. 5.3.

The argument given just now for the occurrence of the collective oscillation should be considered *cum grano salis*. Strictly speaking, there must always be at least one other particle within a radius of λ_s otherwise no screening could emerge. Note that the paradoxical situation arises from the simultaneous use of a continuous model with charge density $\tilde{\rho}(\vec{r})$ and a discrete model (single electrons).

In the shaded region of Fig. 5.3, collective phenomena will not play a dominant role. However, plasmons are defined for all densities, as we have seen. Since λ_s has been calculated statically, our result only implies that in the non-shaded region collective oscillations are induced even by slow disturbances, whereas in the remaining part of the diagram a sufficiently quick (dynamic) disturbance is necessary. This is the reason why plasmons are not excited in a metal under "normal" conditions.

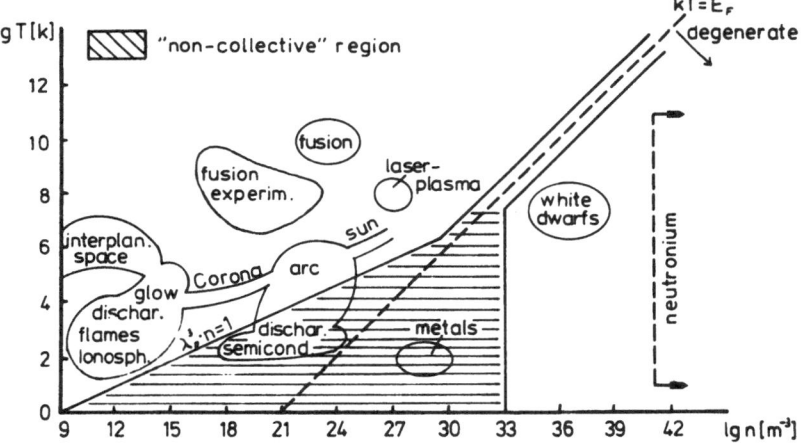

Fig. 5.3. Collective and non-collective behaviour of plasmas (schematic).

5.3. Plasmon Dispersion

In the following, we shall examine the longitudinal oscillation more closely than in Sect. 4.7.

As we have seen, the conduction electrons in a medium oscillate at frequency ω_p (where $\varepsilon = 0$), when damping is neglected—see Eq. (4.33). This mode is longitudinal, i.e. $\vec{k}\|\vec{E}$. According to Eq. (4.14c),

$$\nabla \times \tilde{\vec{E}} = -\frac{1}{c}\frac{\partial}{\partial t}\tilde{\vec{B}} \leftrightarrow i(\vec{k} \times \vec{E}) = \frac{i\omega}{c}\vec{B} \qquad (4.14c)$$

there is no magnetic field, hence the equation of motion (4.28) did not contain any Lorentz force. From (4.26), we have

$$\vec{j}_p = -i\omega\frac{\varepsilon - 1}{4\pi}\vec{E} \qquad (5.7)$$

which means that the electrons oscillate parallel to $\vec{k}$, $(\vec{j}\|\vec{k})$.

From the continuity equation for the polarization current

$$0 = \nabla\tilde{\vec{j}}_p + \frac{\partial\rho_p}{\partial t} \longleftrightarrow (\vec{k},\vec{j}) = \omega\rho_p \qquad (5.8)$$

it follows that

$$\rho_p = -i\frac{\varepsilon - 1}{4\pi}(\vec{k},\vec{E}) \qquad (5.9)$$

does *not* vanish, in general. (It does so only in the limit $\omega \to \infty$, since $\lim_{\omega\to\infty}\varepsilon = 1$). The polarization charge density has the same spatial periodicity as the longitudinal component of $\vec{E}$.

In the case that $\rho_p \neq 0$, we have density fluctuations and one should consequently expect diffusion mechanisms to take place. This can be considered by an additional diffusion term in Eq. (4.28):

$$m\frac{\partial\tilde{\vec{v}}}{\partial t} + \frac{\nabla p}{n_0} = e\tilde{\vec{E}} \longleftrightarrow -i\omega m\frac{\vec{j}_p}{\rho_0} + i\vec{k}\frac{p}{n_0} = e\vec{E}. \qquad (5.10)$$

p is the pressure in the electron gas, and friction is neglected in this context. The equilibrium $(k = 0)$ charge density ρ_0 of the conduction electrons relates to the density as $\rho_0 = n_0 e$. The gradient ∇p

can be written as

$$\nabla p = \left.\frac{\partial p}{\partial n}\right|_0 \cdot \nabla n = \left.\frac{\partial p}{\partial n}\right|_0 \frac{\nabla \rho}{e} \leftrightarrow i\vec{k}p = \left.\frac{\partial p}{\partial n}\right|_0 \frac{i\vec{k}\rho_p}{e}. \tag{5.11}$$

Now we use Eq. (5.8) to express ρ_p through j_p, and (5.7) to eliminate $\vec{E}$ from (5.10). Then, for the longitudinal components, we have

$$-i\frac{\omega m}{n_0 e}j_p + i\frac{1}{n_0 e}\left.\frac{\partial p}{\partial n}\right|_0 k^2 \frac{j_p}{\omega} = i\frac{4\pi e}{\omega(\varepsilon - 1)}j_p. \tag{5.12}$$

Eq. (5.12) has non-trivial solutions if and only if

$$\omega^2 = \omega_c^2 = \frac{4\pi n_0 e^2}{m} + \frac{1}{m}\left.\frac{\partial p}{\partial n}\right|_0 k^2. \tag{5.13}$$

The first term is the square of the plasma frequency ω_p^2 derived previously. Inclusion of the diffusion term causes the eigenfrequency of the longitudinal collective oscillation to depend on k. Since the latter is defined at $\varepsilon(k, \omega) = 0$, Eq. (5.13) is equivalent to

$$\varepsilon(k, \omega_c(k)) = 0. \tag{5.14}$$

$\partial p/\partial n$ can be calculated from the adiabatic state equation of the electron gas, (since no thermal equilibrium is achieved for the quick movements of the electrons)

$$\frac{p}{p_0} = \left(\frac{n}{n_0}\right)^\gamma \tag{5.15}$$

where

$$\gamma = \frac{c_p}{c_v} = \frac{F + 2}{F} \tag{5.16}$$

and F is the number of degrees of freedom. Since the charges oscillate in *one* direction ($\vec{k}$), $F = 1$ [4.8]. Here we did assume that dipole polarizability is negligible in the frequency domain considered. Otherwise we would have $F \neq 1$ because of rotational freedom.

So

$$\left.\frac{\partial p}{\partial n}\right|_0 = 3\frac{p_0}{n_0}. \tag{5.17}$$

The zero point pressure p_0 of a degenerate electron gas in terms of the Fermi energy E_F is approximately [4.13]

$$p_0 = \frac{2}{3}n_0\frac{3}{5}E_F, \tag{5.18}$$

hence

$$\left.\frac{\partial p}{\partial n}\right|_0 = \frac{6}{5}E_F = \frac{3}{5}mv_F^2 \tag{5.19}$$

and the frequency dispersion of the collective oscillation (5.13) is

$$\omega_c^2 = \omega_p^2 + \frac{3}{5}v_F^2 k^2. \tag{5.20}$$

Contrary to the dispersion Eq. (4.33)—where diffusion was neglected — the group velocity no longer vanishes now. The collective mode is capable of transporting energy.

5.4. Boundaries

Thus far we have dealt with unbounded media, the eigenmodes of which are described by the homogeneous Eq. (4.23). The solutions we found are eigenfunctions of the infinite, unbounded medium. As such, they are determined by the material properties. When the medium has a boundary, our basic assumption of homogeneity no longer holds, and different wave vector components of the fields are coupled. The matrix equation (4.22) is an integral equation now, containing all Fourier coefficients in k-representation at once. The coupling prevents a straightforward solution.

It is convenient to treat the influence of boundaries in a more transparent way, using such facts as that the normal component of $\vec{D}$ is continuous across the boundary. This approach, then, yields the well-known Fresnel equations, which, together with the laws of reflection and refraction, allow for a complete determination of the fields inside the bounded medium when the fields outside are given. The following discussion deals with the response of a medium possessing a dispersion relation like that in Fig. 4.5 when its (flat) surface is exposed to an external field. In doing so, we shall treat the problem qualitatively, without resorting to the Fresnel formulas.

Having in mind the comprehensive topics of electron scattering—to be dealt with in Sect. 5.8 and 5.9—we refer to a medium of slab geometry. The geometry of the boundary and the respective components of the wave vector are depicted in Fig. 5.4.

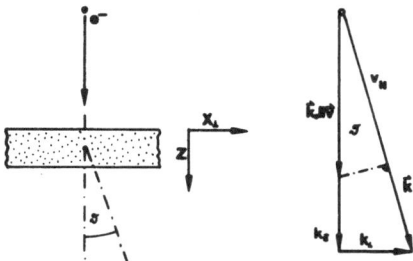

Fig. 5.4. Slab geometry of the medium (left) and notation of components of the wave vector $\vec{k}$ and the velocity $\vec{v}$ of an incident electron.

We shall find it useful to give dispersion relations in $(k_\perp, \omega)$-diagrams. Regions in the $(k_\perp, \omega)$-plane can be separated as to their optical behaviour. For a medium with one oscillator at ω_t (cf.Eq. (4.35)) six domains can be discerned. They are bordered by the light line $\omega = k_\perp c$, the line $\omega = \omega_t$, and the transverse dispersion $k_t(\omega) = \sqrt{\varepsilon}\,\omega/c = \omega/v_{ph}$ ($c/\sqrt{\varepsilon} = c/n$ is the phase velocity in the medium) (Fig. 5.5).

Obviously, transverse waves in I and II can penetrate both the medium and the vacuum, since

$$k_z = \sqrt{k_t(\omega) - k_\perp^2} \qquad in \qquad I \qquad \epsilon R,$$
$$k_z = \sqrt{\omega/c - k_\perp^2} \qquad in \qquad II \qquad \epsilon R. \tag{5.21}$$

In III, k_z is imaginary in the medium, but real in vacuum; this is the region of total reflection of waves coming from vacuum (e.g. radio waves totally reflected at the ionosphere).

In IV, k_z is real in the medium, but imaginary in vacuum—here waves cannot leave the medium (e.g. light guides).

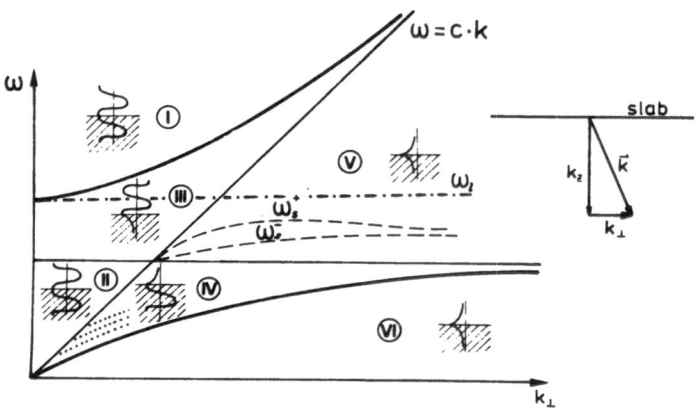

Fig. 5.5. Regions I-VI in a $(k_\perp, \omega)$-diagram correspond to different optical behaviour. For explanation see text.

In V, VI, k_z is imaginary both in vacuum and the medium—these waves cannot leave the boundaries.

It should be added that in thin slabs, standing wave fields evolve due to the boundary conditions. As is known from eigenvalue problems in optics or quantum mechanics, each standing wave corresponds to a particular discrete k. In the $(k_\perp, \omega)$-diagram these modes are drawn schematically (dotted lines in Fig. 5.5). In addition, we have drawn the surface plasmon (see later), which splits into two branches (dashed) for similar reasons as there are more than one light modes in a bounded medium. In principle, there should also be a splitting of the longitudinal mode ω_l (dash-dotted in Fig. 5.5)—however, its slope is so shallow that all branches practically coincide.

5.5. Surface Oscillations

As we have seen in the preceding section, waves are bound to the surface in regions V and VI of the $(k_\perp, \omega)$-diagram. Still it is not yet clear whether or not an eigenmode exists in these regions. Consider the electric field near the boundary as $\vec{E} = -\nabla\varphi$ with

$$\varphi(\vec{x}, z) = e^{i\vec{k}_\perp \vec{x}} \cdot e^{-k_z \cdot |z|} \qquad (5.22)$$

where z is the coordinate perpendicular to the surface defined by $\vec{r} = (\vec{x}, z = 0)$. Note that (5.22) and $\vec{E} = -\nabla\varphi$ is an *ansatz* neglecting retardation effects. From the first of Maxwell's equations (4.14a) we know that the normal component of $\vec{D}$ across the surface is continuous, which yields immediately, taking the normal derivatives of the potential (5.22)

$$k_z \varphi(\vec{x}, 0) = -k_z \varphi(\vec{x}, 0) \cdot \varepsilon \qquad (5.23)$$

or

$$\varepsilon = -1. \qquad (5.24)$$

A surface oscillation can only be supported at frequencies and wave vectors where (5.24) holds. (In case of the free electron gas, Eq. (4.34), the surface oscillation (surface plasmon) is resonant at $\omega_s = \omega_p/\sqrt{2}$). Because of hybridisation between the surface mode and the photon there will be a change in the dispersion relation (5.24) which was derived under neglect of retardation $(c \to \infty)$ or, what is equivalent, $k \gg \omega/c$ [5.6]. The new ("retarded") mode is drawn as a dashed line in Fig. 5.5. Near the light line where the graph bends, the mode is coined "surface plasmon polariton".

The reader interested in details of surface plasmons may wish to consult the monograph of Raether [5.16].

5.6. The ATR-Method

Surface plasmons provide indirect information on the plasma frequency of the bulk. So, if it were possible to excite surface oscillations by light, one would have at hand a simple method of investigation. However, when light is incident on a slab surrounded by vacuum or any medium, surface oscillations cannot be excited. This is because $k_\perp$ is always left to the light line, whereas the surface plasmon dispersion is defined at higher $k_\perp$ (Fig. 5.6).

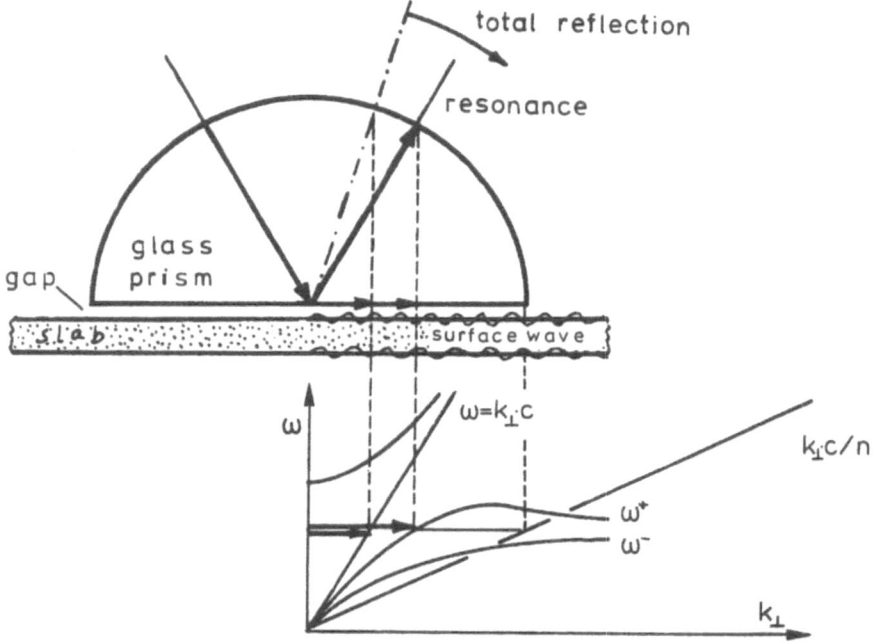

Fig. 5.6. Attenuated total reflection (ATR)-geometry for excitation of surface modes.

A tricky method to circumvent this seeming impossibility has been contrived in the last decade [5.10]. A prism or better a half cylinder of glass is placed immediately above the slab, leaving a small distance between the two surfaces. Thus, $k_\perp$ of the electromagnetic wave extends up to the light line of the prism, depending on the angle of incidence. Matching of $k_\perp$ to the surface plasmon dispersions $\omega^{+,-}$ is possible between the vacuum light line and the prism light line, i.e. when the incidence angle is in the region of total reflection. Since the fields decay exponentially from the prism surface into vacuum, they penetrate the slab only if the gap between glass and slab is sufficiently small (so to speak, they couple to the slab).

At those angles of incidence where $k_\perp$ matches $\omega^{+,-}$, energy is transferred from the beam to the surface oscillation, which causes a marked attenuation of the totally reflected beam. Therefore, the method is termed attenuated total reflection (ATR) or prism method.

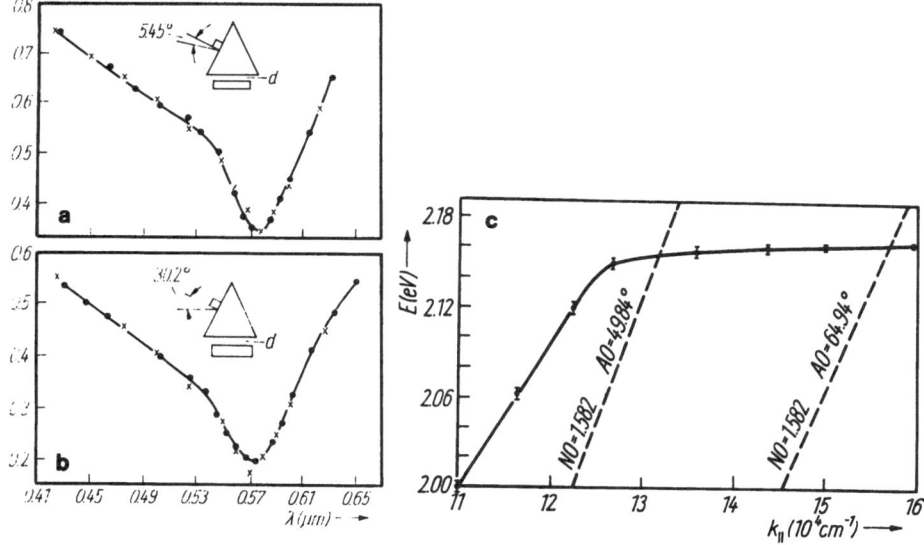

Fig. 5.7. a,b) Attenuated total internal reflection spectrum of copper plotted as the normalized value of the reflectivity versus wavelength. The angle of incidence of the incoming light with respect to the input side of the light flint prism is shown in the ATR diagram in the insert. × computer, • experimental. c) Dispersion curve of the surface plasmon of copper (solid line) derived from ATR-minima for various angles of incidence. Also shown is the energy-wave vector light line generated by the prism for the experimental spectra of a) and b) where the low wave vector line corresponds to a) and the high wave vector line to b). The indicated angle (AO) is with respect to the normal of the prism base. From [5.5].

Thus the surface oscillation is optically accessible. Some results of ATR-measurements are shown in Figs. 5.7 and 5.8. It is worthwhile noting that recently the acronym ATR has also been used for investigation of bulk modes. By firmly placing the glass prism or cylinder onto the material to be investigated, one has access to the region left of the glass light line $\omega = k \cdot c/n$ in Fig. 5.6, by choosing the frequency and the angle of incidence.

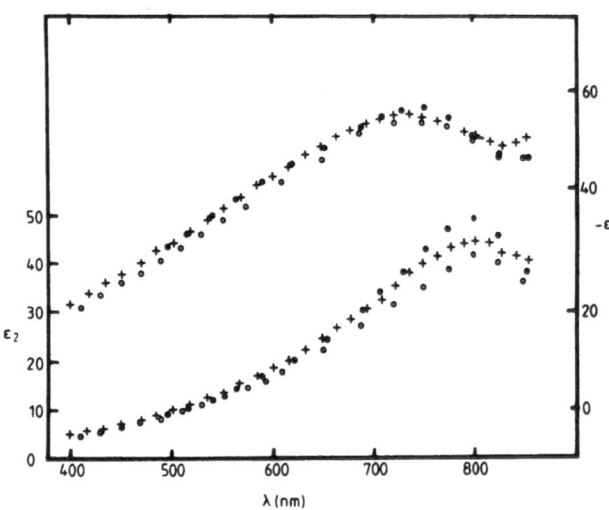

Fig. 5.8. Real and imaginary parts of the Al dielectric function $\varepsilon = \varepsilon_1 + i\varepsilon_2$ determined: from reflectivity measurements with surface plasmons excitation $(+)$; from optical reflectivity and transmissivity measurements $(\circ)$. The values of Mathewson and Myers [5.9] are also indicated $(\bullet)$. From [5.8].

Of course, the $\omega^\pm$-modes which were defined at the medium-vacuum interface, are no longer existent and have been replaced by new surface modes at the medium-glass interface. These are located to the right of the glass light line, hence cannot be excited now. But in this setup, *bulk* modes of the medium, such as longitudinal and transverse phonons, show up just as the surface modes did in the "genuine" ATR-geometry.

The method has been applied successfully to the determination of phonon dispersion—see for instance [4.17] and Fig. 4.9.

5.7. On the Restricted Validity of the Fresnel Equations

Calculation of wave fields in a medium is performed by the Fresnel Eqs. in optics, i.e. the *homogeneous* solutions discussed above are fitted to the vacuum wave.

This is done using continuity conditions of the fields at the boundary. They are obtained by a "pill-box" integration from the Maxwell equations [4.8]. In case of p-polarized light ($\vec{E}$ is parallel to the plane of incidence) we have continuity of the tangential component of $\vec{E}$:

$$E_{0x} - E_{rx} = E_{tx} \tag{5.25}$$

and continuity of the normal component of $\vec{D}$

$$\varepsilon_0(E_{0z} + E_{rz}) = \varepsilon_1 E_{tz}. \tag{5.26}$$

Snell's law allows replacing the four unknown components of $\vec{E}_t$, $\vec{E}_r$ by their moduli, which are then uniquely determined by Eqs. (5.25), (5.26). This standard derivation implies that both $\vec{E}$ and the polarization current (4.12c) are discontinuous at the boundary, as can be seen from inserting Eq. (4.26) in the expression for $\vec{D}$:

$$\vec{D} = \varepsilon \cdot \vec{E} = \vec{E} + \frac{4\pi i}{\omega} \cdot \vec{j} \tag{5.27}$$

since the normal component of the right side of Eq. (5.27) is continuous, according to Eq. (5.26). A discontinuous current implies an infinite surface charge, which helps to avoid any charge density oscillations within the medium [5.12]. Apart from the unphysical demand of singular surface charges, we have seen that charge oscillations in a medium may well exist — in metals they are even the rule, not the exception.

Inspection of Fig. 5.5 shows that at frequencies slightly above ω_p two modes may be excited in the medium: a transverse one at $k_t(\omega)$ and a longitudinal one at $k_l(\omega)$. So two electric field vectors are superimposed in the medium: $\vec{E}_t$, $\vec{E}_l$, each corresponding to the respective k (see Fig. 5.9). Consequently, we have $\vec{j}_{l,t} = \sigma_{l,t} \cdot \vec{E}_{l,t}$, and $\sigma_{l,t}$ are the longitudinal (transverse) conductivities.

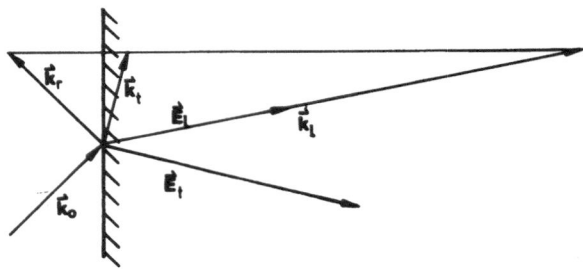

Fig. 5.9. Excitation of transverse (E_t) and longitudinal (E_l) bulk modes by electromagnetic radiation at a frequency close to ω_p.

It is then justified [5.13] to replace condition (5.26) by

$$E_{0z} + E_{rz} = E_{tz} + E_{lz} \tag{5.28}$$

and

$$j_z = 0 = \sigma_t E_{tz} + \sigma_l E_{lz}. \tag{5.29}$$

The latter holds because σ is a tensor [5.12]. Contrary to the Fresnel case, we have now to solve for *three* vectors $\vec{E}_r, \vec{E}_t, \vec{E}_l$ (i.e. six scalar unknowns). The solution is achieved by the three Eqs. (5.25), (5.28), (5.29) together with the three conditions for the ratio of E_x, E_z, known as law of reflection and Snell's law, for E_t and E_l respectively. See Fig. 5.9 for the relative orientation of the electric fields. There is now a longitudinal component of $\vec{E}$ in the medium. It seems perhaps paradoxical that a transverse field can excite a longitudinal mode. This is possible because the surface of the medium breaks rotational symmetry, thus σ becomes a tensor, and coupling between longitudinal currents and transverse electric fields is no longer forbidden.

The occurrence of an additional (longitudinal) mode should change the reflection coefficient as compared to the Fresnel value since in derivation of the Fresnel equations the possibility of longitudinal modes is ignored. A detailed calculation [5.12] however, shows that the change is effective only in the close vicinity of ω_p since elsewhere the plasmon does not exist. Although vacuum cannot support a longitudinal wave, and although there is no direct coupling between the transverse photon and the longitudinal plasmon, the latter is excited when the medium is hit by photons of a frequency slightly above ω_p at an angle close to the limit of total reflection. This is because the existence of a surface breaks translational symmetry. As already mentioned, under these circumstances the different Fourier components which are decoupled in Eq. (4.22) remain coupled, allowing for excitation of plasmons by a photon. See Fig. 5.10 for the effect of plasmon excitation on reflectivity of metals.

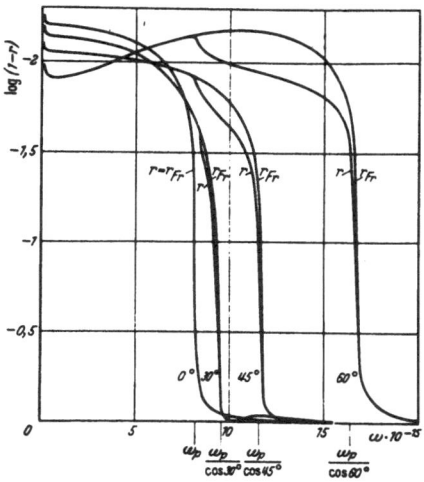

Fig. 5.10. The reflection coefficient r, calculated by nonlocal optics for different angles of incidence for Na; r_{Fr} from local optics (Fresnel equations). From [5.12].

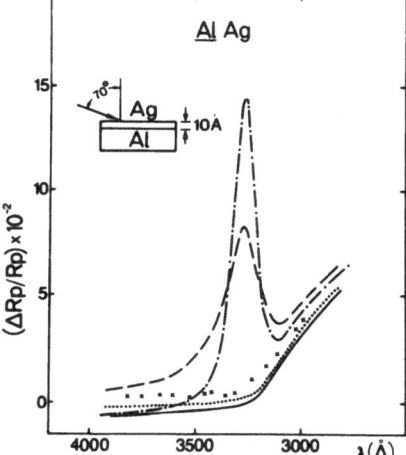

Fig. 5.11. Relative reflectance difference 2 [R (free)— R (covered)]/[R (free)+R (covered)] for Al covered by an Ag film of 10 Å.
xxxx measurements, — · — · — standard optics, - - - - damping increased by factor 10, — nonlocal optics, ... damping increased by factor 10. From [5.12].

Fig. 5.11 shows measurements of the reflectivity of thin Ag films on Al as a function of light frequency. The experimentally missing peak is strong evidence for the correctness of the arguments given above. However, those ideas have not been accepted generally and are still debated [5.6], [5.14], [5.17].

5.8. The Differential Cross Section

The (power) dissipation density in a medium of volume V is given by

$$\tilde{L}(\vec{r},t) = \vec{\tilde{E}}(\vec{r},t)\vec{\tilde{j}}(\vec{r},t) \qquad (5.30)$$

where $\vec{j}$ is the *external* current density, and $\vec{E}$ the field in the medium (Ohmic power $P = U \cdot I = \vec{E} \cdot V \cdot \vec{j}$, $L = \partial P/\partial V = \vec{E} \cdot \vec{j}$). The total energy lost by the external charge carrier is

$$\int dt \int d^3\vec{r} \tilde{L}(\vec{r},t) = W. \qquad (5.31)$$

We consider a probe electron which has entered a slab at normal incidence in a coordinate system as drawn in Fig. 5.12.

The energy lost by the electron per unit path length is

$$w = \partial W/\partial z = \int d^2 x_\perp \int dt \tilde{L}(\vec{r},t). \qquad (5.32)$$

We shall again apply Fourier transforms, but only in the $x_\perp$-plane in this case, due to the finite extension of the slab in z, leaving z as a free variable. Accordingly, we denote the transform L_z

$$\tilde{L} \leftrightarrow L_z(\vec{k}_\perp,\omega) := \iint \tilde{L}(z,\vec{x}_\perp,t)e^{-i(\vec{k}_\perp,\vec{x}_\perp-\omega t)} d^2 x_\perp dt \qquad (5.33)$$

bearing in mind that z is a parameter of $\tilde{L}$.

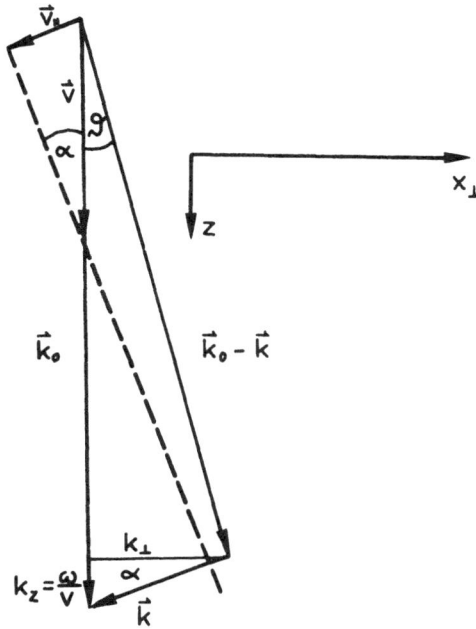

Fig. 5.12. Scattering geometry for perpendicular incidence of a probe particle onto a slab.

From Eqs. (5.32), (5.33)

$$w = L_z(0,0). \qquad (5.34)$$

In $(\vec{k}_\perp, \omega)$-representation Eq. (5.30) is a convolution (cf. 4.6b)

$$L_z = \frac{1}{(2\pi)^3} \vec{E}_z * \vec{j}_z. \qquad (5.35)$$

Remember that z is a parameter, and from Eqs. (5.34), (5.35)

$$w = \frac{1}{(2\pi)^3} \int d^2 k_\perp \int_{-\infty}^{+\infty} d\omega \, E_z(-\vec{k}_\perp, -\omega) j_z(\vec{k}_\perp, \omega). \qquad (5.36)$$

Since $\vec{E}$, $\vec{j}$ are real quantities, their transforms obey

$$\begin{aligned} E_z(-\vec{k}_\perp, -\omega) &= E_z^*(\vec{k}_\perp, \omega), \\ j_z(-\vec{k}_\perp, -\omega) &= j_z^*(\vec{k}_\perp, \omega). \end{aligned} \qquad (5.37)$$

In Eq. (5.36) $\int d\omega$ can be split, and using (5.37) we obtain

$$w = \frac{1}{(2\pi)^3} \int d^2k_\perp (\int\limits_0^\infty d\omega E_z^*(\vec{k}_\perp,\omega) j_z(\vec{k}_\perp,\omega) +$$

$$+ \int\limits_0^\infty d\omega' E_z(-\vec{k}_\perp,\omega') j_z^*(-\vec{k}_\perp,\omega')) = \qquad (5.38)$$

$$= \frac{1}{(2\pi)^3} \int d^2k_\perp \int\limits_0^\infty d\omega (E_z^* j_z + E_z j_z^*).$$

On the other hand, w can be written in terms of the differential scattering probability $\partial^3 P/\partial^2 k_\perp \partial\omega$ as

$$w = \int d^2k_\perp \int\limits_0^\infty d\omega \underbrace{\hbar\omega}_{energy} \underbrace{\partial^3 P/\partial^2 k_\perp \partial\omega}_{probability\ density} . \qquad (5.39)$$

P has dimension "probability/unit path length".
    Comparing (5.38), (5.39),

$$\partial^3 P/\partial^2 k_\perp \partial\omega = \frac{1}{(2\pi)^3 \hbar\omega} (j_z^* E_z + j_z E_z^*). \qquad (5.40)$$

The differential scattering probability (5.40) will exhibit two maxima in general, corresponding to longitudinal and transverse excitations. The latter turns out to excite Cerenkov radiation which is strong at large ε_1 and small ε_2. We shall not dwell on Cerenkov radiation which is negligible for most materials and electron velocities, but rather examine the longitudinal part. Since the medium was assumed isotropic, $\vec{j} = \sigma\vec{E}\|\vec{E}$, and the longitudinal part of (5.40) is

$$\partial^3 P_\|/\partial^2 k_\perp \partial\omega = \frac{1}{(2\pi)^3 \hbar\omega} (j_{z_\|}^* E_{z_\|} + j_{z_\|} E_{z_\|}^*). \qquad (5.41)$$

5.9. Energy Loss Function

In case of a single probe electron of velocity $\vec{v}$ at normal incidence (see Fig. 5.12)

$$\tilde{j}_{\parallel} = v_{\parallel} e \delta(\vec{r} - \vec{v}t) \longleftrightarrow j_{\parallel} = 2\pi e v_{\parallel} \delta(\omega - k_z v) \tag{5.42}$$

and

$$\dot{j}_{z_{\parallel}}(\vec{k}_{\perp}, \omega) = \frac{1}{2\pi} \int j_{\parallel} e^{ik_z z} dk_z. \tag{5.43}$$

From Fig. 5.12

$$v_{\parallel}/v = k_z/k = k_z \sqrt{k_z^2 + \vec{k}_{\perp}^2} \tag{5.44}$$

and

$$\dot{j}_{z_{\parallel}} = e \frac{\omega}{v\sqrt{(\omega/v)^2 + k_{\perp}^2}} e^{i\frac{\omega z}{v}}. \tag{5.45}$$

$E_{\parallel}$ is obtained from the first Maxwell Eq. (4.14a) and the transform of the δ-function in (5.42):

$$\nabla \tilde{D} = 4\pi \tilde{\rho} = 4\pi e \delta(\vec{r} - \vec{v}t) \leftrightarrow i\vec{k}\varepsilon\vec{E} = 8\pi^2 e \delta(\omega - k_z v), \tag{5.46}$$

and

$$E_{z_{\parallel}} = \frac{1}{2\pi} \int dk_z E_{\parallel} e^{ik_z z} = \frac{4\pi e}{ivk'} \frac{e^{i\omega z/v}}{\varepsilon(k', \omega)} \tag{5.47}$$

where $k'^2 = (\omega/v)^2 + k_{\perp}^2$. Eq. (5.41) becomes

$$\partial^3 P_{\parallel}/\partial^2 k_{\perp} \partial\omega = \frac{1}{(2\pi)^3 \hbar\omega} \frac{4\pi e^2 \omega}{v^2 k'^2} \left(\frac{1}{i\varepsilon} - \frac{1}{i\varepsilon^*} \right). \tag{5.48}$$

The bracket can be replaced by $2 \cdot Im(1/\varepsilon)$ and from Fig. 5.12, we have

$$d\omega d^2 k_{\perp} = k_0^2 \cos\vartheta d\Omega dE/\hbar. \tag{5.49}$$

For fast electrons, (i.e. primary energy $E_0 \gg$ energy loss E) the scattering probability decreases quickly with $k_{\perp}$.

Hence, in this case, we have mainly forward scattering, $\cos\vartheta \approx 1$, and we may rewrite Eq. (5.48) as

$$\partial^3 P_{\parallel}/\partial^2 \Omega \partial E = \frac{8\pi e^2 k_0^2}{(2\pi)^3 v^2 k'^2 \hbar^2} Im\left(\frac{1}{\varepsilon(k',\omega)}\right) =$$
$$= Im\left(\frac{1}{\varepsilon}\right)/(e\pi a_0)^2 k'^2. \tag{5.50}$$

Eq. (5.50) is the basic formula for interpretation of energy loss spectra in the low and medium energy range. Once $Im(1/\varepsilon)$ is known, Kramers-Kronig-analysis (KKA) yields $Re(1/\varepsilon)$, and, eventually, $\varepsilon(k,\omega)$ can be derived. Fig. 5.13 exemplifies these calculations.

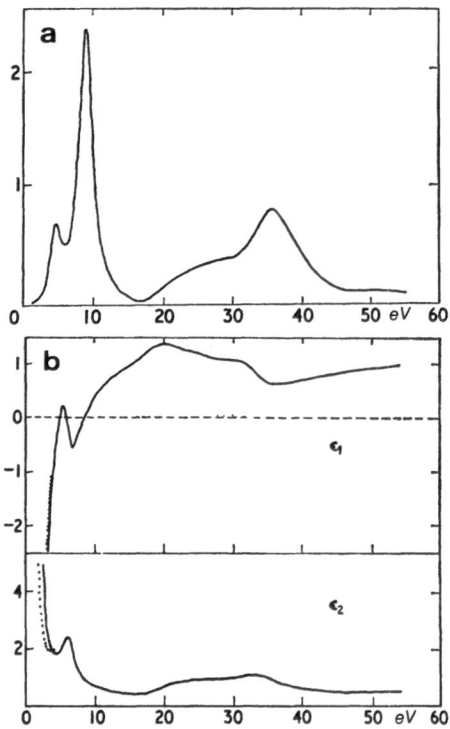

Fig. 5.13. a) Energy loss function $Im(1/\varepsilon)$ of Ca and b) dielectric function $\varepsilon = \varepsilon_1 + i\varepsilon_2$ calculated by Kramers-Kronig-analysis from a). The oscillator which shows up in ε_2 at ~ 6 eV is an interband transition. Results from optical measurements [5.9] are shown for comparison as dotted lines. From [5.15].

These examples demonstrate two advantages in using ELS instead of optical methods: ELS covers a much larger energy range than optics, so not only $Im(1/\varepsilon)$ contains more information, but also KKA can be performed much more accurately, and b) by angle-resolved measurements one has access to the k-dependent dielectric function, whereas optical methods are, in principle, restricted to $\varepsilon(0, \omega)$. From these considerations, it is evident that inelastic electron scattering is of similar importance for probing electronic excitations as is neutron scattering for the investigation of phonons.

6. Quantum Mechanical Preliminaries

Before going into details of the quantum mechanical description of the electron gas as a many-body problem we briefly digress on the basic concepts of perturbation theory for solving the Schrödinger equation.

Green functions (or Green operators) are a powerful tool in reaching this aim, especially in many-electron problems. We shall outline only the basic principles necessary for a fundamental understanding. The present chapter is arranged in a self-contained, albeit very condensed form. The reader should be familiar with fundamental definitions of quantum theory, such as wave functions and their meaning, the Schrödinger equation, operators, Dirac's notation of vectors, and the like. It is also useful to know about fundamentals of the theory of complex variables, Fourier transforms and linear algebra. Beginners or those interested are advised to have a look into one of the standard textbooks on these topics, for a more precise treatment of what will be discussed on the following pages, the excellent textbooks of R.D. Mattuck [6.4] and S. Raimes [6.3] are recommended. The monograph of Economou [3.17] is also very useful.

In the following sections, up to discussion of second quantization, we deal solely with a one-electron system. Note that the treatment of a many-electron system under neglect of any interaction is similar to that of the one-particle formalism. The reason is that—as we shall see in the next chapter— the energy of a non-interacting system of electrons is the sum over allowed single-electron energies, and the wave function can be written as an antisymmetrized product of single-electron wave functions. This is due to the separability of the Schrödinger equation in the absence of interactions. When interaction is present, this is no longer the case. As a consequence, the energy of the system is changed with respect to the non-interacting system, and the wave function can no longer be written as an antisymmetrized product of single-particle wave functions.

However, the use of Green functions (or Green operators) remains essentially valid in the form that will be described in the next sections.

6.1. Summary of Important Facts

At the beginning of this fatiguing chapter, we shall list the most important results. The list is intended as an incentive to continue for the reader who wants to skip the chapter for any reason whatsoever. (Most probably, he is not willing to attend a pseudo-philosophical kindergarten course where understanding like physics seems to be a matter of chance.)

— The Schrödinger equation for a (one-electron) system with perturbation W can be solved by means of the Green operator

$$G(\omega) = (\hbar\omega - H)^{-1} = (\hbar\omega - H_0 - W)^{-1} \qquad (6.15)$$

as

$$|\varphi> = |\vec{k}_0> + GW|\vec{k}_0> \qquad (6.21)$$

where $|\vec{k}_0>$ is a free particle eigenstate.
— The Green operator G can be found from the series expansion

$$G(\omega) = \sum_{n=0}^{\infty} (G_0(\omega) \cdot W)^n \cdot G_0(\omega) \qquad (6.19)$$

with respect to the unperturbed Green operator G_0.
— The poles of $G_0(\omega)$ $(G(\omega))$ define the resonant frequencies of the system described by H_0 (H)
— $G_0(\omega)$ is diagonal in k-representation:

$$G_0(\vec{k}_i, \vec{k}_j, \omega) = <\vec{k}_i|G_0(\omega)|\vec{k}_j> = \frac{\delta_{ij}}{\hbar(\omega - \omega_j)}. \qquad (6.14)$$

— The Fourier transform of a Green operator $G(\omega)$ with respect to time describes "propagation" of states trough time. Wavefunctions at different times are related as

$$\varphi(\vec{k}_j, t) = i\hbar \sum_i G_{00}(\vec{k}_j, \vec{k}_i, t)\varphi(\vec{k}_i, 0). \qquad (6.33)$$

Therefore, "propagator" is a synonym for Green operator.

— In second quantization, one- and two-particle operators A, B are given in terms of creation and destruction operators c_i^+, c_j as

$$A \hat{=} \sum_{ij} A_{ij} c_i^+ c_j, \tag{6.44}$$

$$B \hat{=} \sum_{ijkl} B_{ijkl} c_j^+ c_i^+ c_k c_l \tag{6.45}$$

with

$$\begin{aligned} A_{ij} &= < k_i|A|k_j > \\ B_{ijkl} &= < k_i k_j|B|k_k k_l > . \end{aligned} \tag{6.46}$$

— The matrix elements of the propagator G_0 are probability amplitudes for creating and destroying states in a non-interacting system

$$G_0^+(k, k', t) \hat{=} \frac{i}{\hbar} e^{-i\omega_0 t} < 0| \quad c_k(t) \quad c_{k'}^+(0)|0 > \Theta(t). \tag{6.55b}$$

In case of interactions, G_0 is replaced by G.

— Feynman graphs are pictorial representations of scattering events contributing to the probability amplitude for propagation of a state. Graphical expressions can be manipulated like algebraic expressions (Addition and multiplication are defined for graphs).

— the perturbation series for G can be much simplified by the technique of selective summation: that is picking out those graphs (events) which are thought to contribute a great deal to the propagation probability.

— The Coulomb interaction between two electrons is

$$\hat{=} \langle k+q \; l-q| V|k l\rangle \tag{6.66}$$

$$V_{k+q,l-q,k,l} = V_q = \frac{4\pi e^2}{q^2} \tag{6.67}$$

6.2. The Lippman-Schwinger Equation

The time-independent Schrödinger equation for the wave function $\varphi(\vec{r})$ of a single electron of mass m moving under the influence of a perturbation W is

$$\left(-\frac{1}{2m}\nabla^2 + W(\vec{r})\right)\varphi(\vec{r}) = E\varphi(\vec{r}). \tag{6.1a}$$

W may be thought of as an electrostatic potential. Eq. (6.1a) is equivalent to

$$H|\varphi> = (H_0 + W)|\varphi> = E|\varphi> \tag{6.1b}$$

where $H_0 = P^2/2m$ is the kinetic energy operator (the free particle Hamiltonian). W is the potential energy operator, E is the energy eigenvalue and $|\varphi>$ is a vector in Hilbert space, written in the Dirac-notation as a "ket".

It is well known that the solutions of Eq. (6.1b) for *vanishing* perturbation $W \equiv 0$ are free particle eigenstates $|\vec{k}_i>$ (plane waves with wave vector $\vec{k}_i$), and the energy is given by

$$E(\vec{k}_i) = \frac{\hbar^2 k_i^2}{2m} := \hbar\omega_i \tag{6.2}$$

in which we assumed—as we shall in the remainder of the present chapter—that the set of $\vec{k}$ is discrete, i.e. we have to impose boundary conditions on the surface of a base volume. Without restriction of generality, the base volume is taken as unity.

We may now select a particular solution $|\vec{k}_0>$ and look for the change of the ket $|\vec{k}_0>$ under the influence of a perturbation W which we imagine to be continuously increased from zero to its final strength. In doing so, we shall find it convenient to use a frequency scale for the energy $E = \hbar\omega$ and rewrite Eq. (6.1b) as

$$(\hbar\omega - H_0)|\varphi> = W|\varphi>. \tag{6.3}$$

When we abbreviate, for the moment, the right hand side

$$W|\varphi> := |s>, \tag{6.4}$$

and when we define the inverse of $\hbar\omega - H_0$ as $G_0(\omega)$:

$$G_0(\omega) := (\hbar\omega - H_0)^{-1} \qquad (6.5)$$

which means that

$$(\hbar\omega - H_0) \cdot G_0(\omega) = 1, \qquad (6.6)$$

the solution of (6.3) is formally written as

$$|\varphi\rangle = |\vec{k}_0\rangle + G_0(\omega)|s\rangle, \qquad (6.7a)$$

$$|\varphi\rangle = |\vec{k}_0\rangle + G_0(\omega) \cdot W|\varphi\rangle. \qquad (6.7b)$$

Eq. (6.7b) is known as the Lippman-Schwinger equation.

Let us rest for a while in order to comment on Eqs. (6.5) to (6.7). Eq. (6.7) is proved by multiplying by $(\hbar\omega - H_0)$ from the left which yields Eq. (6.3) since the free particle eigenstate $|\vec{k}_0\rangle$ on the right of (6.7) is the solution of the homogeneous differential equation

$$(\hbar\omega - H_0)|\varphi\rangle = 0 \qquad (6.8)$$

as stated in the paragraph following Eq. (6.1).

$G_0|s\rangle$ is a particular solution of Eq. (6.3) as can be proved by use of Eq. (6.6).

The result (6.7) reminds us of a famous theorem from linear algebra: the general solution of an inhomogeneous equation is the sum of the corresponding "homogeneous" solution and one particular solution. We have to remember, however, that the right hand side $|s\rangle$ of (6.5) is not an inhomogeneity since it contains the unknown $|\varphi\rangle$. So, strictly speaking, Eq. (6.3) cannot be denoted *inhomogeneous*, though the similarity is useful as a mnemonic.

6.3. The Green Operator

The operator $G_0(\omega)$ is called the Green operator of the time-independent problem. Note that G_0 depends on a parameter ω. The subscript on G_0 means correspondence to the unperturbed Hamiltonian H_0 of the *free* electron.

Eq. (6.5) can be made more explicit by multiplying by the unity operator

$$1 = \sum_i |\vec{k}_i ><\vec{k}_i|. \tag{6.9}$$

Notice that since $|\vec{k}_i >$ are eigenvectors to H_0, we have, by expansion of the inverse operator with respect to H_0, and using (6.2):

$$(\hbar\omega - H_0)^{-1}|\vec{k}_i >= \frac{1}{\hbar(\omega - \omega_i)}|\vec{k}_i >, \tag{6.10}$$

hence from (6.5), (6.9) and (6.10)

$$G_0(\omega) = \sum_i \frac{|\vec{k}_i ><\vec{k}_i|}{\hbar(\omega - \omega_i)} \tag{6.11}$$

which is well-defined as long as $\omega \neq \omega_i$, otherwise $G_0(\omega)$ has a pole. Since ω_i are the eigenfrequencies of the free electron, we conclude that the poles of $G_0(\omega)$ define the eigenfrequencies (or resonances) of a system which is described by the Hamiltonian H_0.

The Lippman-Schwinger equation (6.7b) is of a type appropriate for a method of solution often encountered in linear algebra and known as the method of successive approximations [6.1]. When the perturbation W is sufficiently faint, there is reason to hope that the second term on the right of (6.7b) is "less important" than $|\varphi >$, loosely speaking. Then one can try an iteration scheme, starting with $|\varphi >\approx |\vec{k}_0 >$ on the right of Eq. (6.7b), and repeatedly substituting the result into the right. Eventually,

$$|\varphi >= |\vec{k}_0 > +G_0(\omega)W|\vec{k}_0 > +(G_0(\omega)W)^2|\vec{k}_0 > +\ldots =$$
$$= \sum_{n=0}^{\infty}(G_0(\omega)W)^n|\vec{k}_0 > . \tag{6.12}$$

The series (6.12) is the Born series for the perturbed state $|\varphi>$.

For sufficiently faint W, the series converges, since the higher the powers of $G_0 W$, the smaller its contribution to the sum. It can be shown that in this case the geometric series of operators in Eq. (6.12) converges—as does the scalar series—to $(1 - G_0 W)^{-1}$, e.g. [6.2]. The solution of the perturbed Schrödinger equation can then be written as

$$|\varphi> = \sum_{n=0}^{\infty} (G_0(\omega)W)^n |\vec{k}_0> = (1 - G_0(\omega)W)^{-1}|\vec{k}_0> . \qquad (6.13)$$

Eq. (6.13) shows clearly that the problem of solving the Schrödinger equation (6.1) falls apart into two different tasks:
- calculation of the "unpertubed" Green operator $G_0(\omega)$ and the eigenvectors of H_0
- calculation of the inverse of the operator $1 - G_0(\omega) \cdot W$. Note that this is usually achieved by the operator series in Eq. (6.12) unless the perturbation W is suitable for a direct calculation of the inverse operator. When, for instance, W is a diagonal matrix in k-representation, the calculation is trivial since G_0 is also diagonal.

This can easily be shown by taking the matrix elements of G_0 in Eq. (6.11). Due to the orthogonality of eigenvectors $<\vec{k}_i|\vec{k}_j> = \delta_{ij}$,

$$G_0(\vec{k}_i, \vec{k}_j, \omega) := <\vec{k}_i|G_0(\omega)|\vec{k}_j> = \frac{\delta_{ij}}{\hbar(\omega - \omega_j)} \qquad (6.14)$$

which is to say: in k-representation, $G_0(\omega)$ is a diagonal matrix, its entries given by (6.14). We note that Green operators, represented in any basis, such as $\{\vec{k}_j\}$ in Eq. (6.14), are called "Green functions".

6.4. The Dyson Equation

In complete analogy to Eq. (6.5), we define a "perturbed" Green operator $G(\omega)$

$$G(\omega) = (\hbar\omega - H)^{-1} = (\hbar\omega - H_0 - W)^{-1}. \qquad (6.15)$$

From Eq. (6.5), we have

$$G(\omega) = (G_0(\omega)^{-1} - W)^{-1}. \qquad (6.16)$$

An equation of this type is often referred to as a Dyson equation.

Keeping in mind that for two operators A, B which do not commute, $(AB)^{-1} = B^{-1}A^{-1}$, the Dyson equation can be transformed to

$$G(\omega) = (1 - G_0(\omega)W)^{-1}G_0(\omega). \qquad (6.17)$$

By means of the identity

$$(1 - G_0(\omega)W)^{-1} = \sum_{n=0}^{\infty}(G_0(\omega)W)^n \qquad (6.18)$$

derived from Eq. (6.13), we obtain from (6.17)

$$G(\omega) = \sum_{n=0}^{\infty}(G_0(\omega) \cdot W)^n \cdot G_0(\omega) \qquad (6.19)$$

and

$$1 + G(\omega) \cdot W = (1 - G_0(\omega)W)^{-1}. \qquad (6.20)$$

Eq. (6.13) is written, by use of (6.20) in terms of the perturbed Green operator G,

$$|\varphi> = |\vec{k}_0 > + G(\omega)W|\vec{k}_0 > . \qquad (6.21)$$

Note the formal similarity to the Lippman-Schwinger equation (6.7b). However, the right hand side no longer contains the unknown $|\varphi >$, instead the "perturbed" Green operator has to be known in order for the solution to be calculated.

6.5. Green Operators in the Time Domain

The physical interpretation of Green operators is facilitated by the use of Fourier transforms with respect to ω. As already discussed in Chapter 4, products of functions in the ω-representation are changed to convolution integrals in the time-domain—see Eq. (4.6). This is also true for operators.

As an example, we give the Fourier transform of Eq. (6.1b), which yields by use of theorem (4.3b), the time dependent Schrödinger equation

$$\left(i\hbar\frac{d}{dt} - H\right)|\varphi> = 0. \tag{6.22}$$

The ket $|\varphi>$ is now a function of time.

Another example is the transform of Eq. (6.6)

$$(i\hbar\frac{d}{dt} - H_0)G_0(t) = \delta(t) \tag{6.23}$$

which is recognized as the differential equation of the time-dependent Green function.

Its solution is given by the Fourier transform of $G_0(\omega)$, Eq. (6.11). In the following derivation, we start with G_0 in k-representation, Eq. (6.14). Since all off-diagonal elements of $G_0(\vec{k}_i, \vec{k}_j, \omega)$ vanish, we confine ourselves to the transform of the diagonal element denoted $G_0(\vec{k}_i, \omega)$.

Since $G_0(\omega)$ is ill defined at the poles $\omega = \omega_i$, the Fourier transform of $G_0(\vec{k}_i, t)$ is obtained as the limit

$$G_0(\vec{k}_i, t) = \lim_{c \to c_0} \frac{1}{2\pi} \int_c d\omega \frac{1}{\hbar(\omega - \omega_i)} e^{-i\omega t} \tag{6.24}$$

where c_0 is the real ω-axis. Depending on how c_0 is approached several G_0s result. There are two useful (and physically meaningful) definitions of G_0, namely the retarded and the advanced Green functions. The retarded Green function G_0^+ is obtained by approaching c_0 from the upper half ω-plane (fig. 6.1). When $t > 0$, c_0

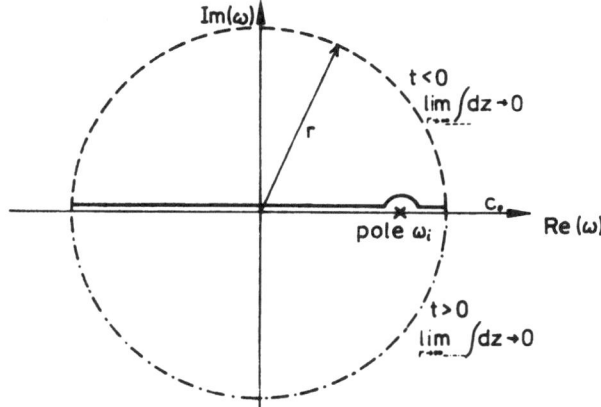

Fig. 6.1. Integration paths for evaluation of the retarded Green's function G.

is closed by a semicircle (of infinite radius) in the lower half plane. Application of the residue theorem yields

$$\int d\omega \frac{1}{\hbar(\omega - \omega_i)} e^{-i\omega t} = -2\pi i Res\left(\frac{e^{-i\omega t}}{\hbar(\omega - \omega_i)}\right) = \frac{-2\pi i}{\hbar} e^{-i\omega_i t}.$$
(6.25)

Since the contribution from the semicircle vanishes, we have

$$G_0^+(\vec{k}_i t) = -\frac{i}{\hbar} e^{-i\omega_i t}, \quad t > 0.$$
(6.26)

When $t < 0$ we close the contour by a semicircle in the upper half ω-plane in order that its contribution to the integral vanishes. In that case, no pole is inside the closed contour, and $G_0^+(t < 0) \equiv 0$. Together with Eq. (6.26)

$$G_0^+(\vec{k}_i, t) = -\Theta(t) \cdot \frac{i}{\hbar} e^{-i\omega_i t}.$$
(6.27a)

The advanced Green function G_0^- is obtained by approaching c_0 from the lower half ω-plane. Applying similar arguments as above,

$$G_0^-(\vec{k}_i, t) = \Theta(-t) \frac{i}{\hbar} e^{-i\omega_i t}.$$
(6.27b)

6.6. Relation of G_0 to the Time Evolution Operator

The time evolution operator U_0 which evolves a ket $|\varphi(t) >$ through time

$$|\varphi(t) >= U_0(t)|\varphi(0) > \qquad (6.28)$$

obeys

$$(i\hbar \frac{d}{dt} - H_0)U_0(t) = 0 \qquad (6.29)$$

as can be seen by "inserting" (6.28) into (6.22) for $W \equiv 0$. The solution of the linear first order differential equation (6.29) is

$$U_0 = e^{-iH_0 t/\hbar}. \qquad (6.30)$$

From the k-representation of (6.30), $U_0(\vec{k}_i, \vec{k}_j, t) = e^{-i\omega_j t} \cdot \delta_{ij}$ and comparing with (6.27), it is plain that

$$U_0(t) = i\hbar \cdot (G_0^+ - G_0^-) := i\hbar G_{00}. \qquad (6.31)$$

In k-representation, Eq. (6.28) reads—using (6.31) and (6.9)—

$$\varphi(\vec{k}_j, t) =< \vec{k}_j|\varphi(t) >$$
$$= i\hbar < \vec{k}_j|G_{00}(t)| \sum_i |\vec{k}_i >< \vec{k}_i|\varphi(0) >, \qquad (6.32)$$

or

$$\varphi(\vec{k}_j, t) = i\hbar \sum_i G_{00}(\vec{k}_j, \vec{k}_i, t)\varphi(\vec{k}_i, 0). \qquad (6.33)$$

Here, the Green function G_{00} reduces to G_0^+ or G_0^-, depending on whether t is positive or negative—see (6.31) and (6.27). Since Eq. (6.33) relates a state evolved through time to the initial wavefunction, it can be said that the retarded Green function "propagates" the wave function from $t_0 = 0$ to t. Therefore, Green functions are synonymously called propagators. (The advanced Green function would propagate states *backwards* in time). The matrix elements of G_0 in any basis $(\vec{k}, \vec{r}, \ldots)$ may thus be viewed as "propagation" probability amplitudes (from $\vec{k}_1$ to $\vec{k}_m, \vec{r}$ to $\vec{r}', \ldots$).

Eq. (6.33) can be used to solve the Schrödinger equation with given initial values $\varphi(\vec{k}_i, 0)$.

Now the question arises as to whether calculations should be performed in the time-domain. In this context Eqs. (6.19), (6.21) are of primary interest.

Since products of operators transform to convolutions when skipping over from ω to time, Eq. (6.19) for the perturbed Green operator contains integral operators $(\int dt)^n$ in the time-domain. These are more difficult to manipulate than the simple products. This fact suggests performing calculations in the frequency-domain.

In closing this section, we mention that the Fourier transform of Eq. (6.7b) is an integral equation with kernel G_0W. The solution, given by the transform of Eq. (6.21), is achieved by the "resolvent kernel" GW, a notation common in the theory of integral equations [6.2]. Therefore, Green operators are sometimes called resolvent operators [6.3].

6.7. Second Quantization

Many-particle-systems are conveniently described in the occupation-number formalism (second quantization). Since we are concerned with electrons, we confine ourselves to fermions. Subsequently the basic features of second quantization for fermions will be reviewed.

The state vector of a *non-interacting* N-body-system is given, after appropriate preparation, by a Slater-determinant

$$|\psi> = \frac{1}{\sqrt{N!}} \begin{vmatrix} |k_1>^{(1)} \ldots\ldots\ldots |k_1>^{(N)} \\ \ldots\ldots\ldots \\ |k_N>^{(1)} \ldots\ldots\ldots |k_N>^{(N)} \end{vmatrix} \qquad (6.34)$$

where the superscript denotes the Hilbert space of the n-th particle, and k_i is the i-th eigenstate of a single particle. For the sake of simplicity, we shall henceforth omit the vector notation in the plane wave states $|\vec{k}_i>$. For short, one can write

$$|\psi> = |k_1 \ldots\ldots k_N> \qquad (6.35)$$

(keeping in mind Eq. (6.34)). Roughly speaking, each state k_i is occupied by *one* particle—since we consider fermions.

Thus, Eq. (6.35) is simplified further by

$$|\psi> = |1_1 \ldots \ldots 1_N> . \tag{6.36}$$

Eq. (6.36) allows us to include the description of less than N particles by writing "0" for each unoccupied *single particle* state, e.g.

$$|\psi> = |1_1 0_2 1_3 \ldots \ldots 1_N> . \tag{6.37a}$$

Generally,

$$|\psi> = |n_1 n_2 n_3 \ldots \ldots n_N> \tag{6.37b}$$

where n_i are occupation numbers 0 or 1 for fermions. This is further simplified by writing out only the occupied states:

$$\psi> = |1_3 1_N> \tag{6.38}$$

for example, describes a system where only two particles are present at states k_3 and k_N. It is worth noting that the "Mickey Mouse" arguments presented above shouldn't be discussed with a theoretician. However, he would, by using Fock spaces, antisymmetrizing operators, orthogonality relations and so on arrive at (6.37) as well.

Now, it is convenient to define operators c_i, c_i^+ by

$$c_i|n_1 \ldots n_i \ldots n_N> = (-1)^{\sigma_i}\sqrt{n_i}|n_1 \ldots n_i - 1 \ldots n_N> \tag{6.39a}$$

$$c_i^+|n_1 \ldots n_i \ldots n_N> = (-1)^{\sigma_i}\sqrt{1 - n_i}|n_1 \ldots n_i + 1 \ldots n_N> \tag{6.39b}$$

where $\sigma_i = n_1 + n_2 + \ldots + n_{i-1}$.

The operator c_i destroys a particle from state k_i, c_i^+ creates such a particle.

Obviously, the expectation value of $c_i^+ c_i$ for $|\psi> = |n_1 \ldots n_i \ldots>$ is, according to Eq. (6.39a)

$$<\psi|c_i^+ c_i|\psi> = n_i <\psi'|\psi'> = n_i, \tag{6.40}$$

the occupation number in state k_i. Hence, the operator

$$\rho_{k_i} = c_i^+ c_i \tag{6.41}$$

is the number density operator. More generally, the $k_i - k_j$-state density operator is

$$\rho_{k_i k_j} = c_i^+ c_j. \tag{6.42}$$

We mention that c_i^+, c_j obey the anticommutation rule

$$[c_i^+ c_j]^+ := c_i^+ c_j + c_j c_i^+ = \delta_{ij}. \tag{6.43}$$

6.8. Operators in Second Quantization

By use of the creation and destruction operators c^+, c other operators can be expressed simply. For instance, for a one-particle operator A we have the equivalence

$$A \hat{=} \sum_{ij} A_{ij} c_i^+ c_j, \tag{6.44}$$

whereas for a two-particle operator B

$$B \hat{=} \sum_{ijkl} B_{ijkl} c_j^+ c_i^+ c_k c_l \tag{6.45}$$

where

$$\begin{aligned} A_{ij} &=< k_i|A|k_j >, \\ B_{ijkl} &=< k_i k_j|B|k_k k_l > . \end{aligned} \tag{6.46}$$

The equivalence stated in Eq. (6.44) is proved as soon as the matrix elements of A are shown to be identical in the "normal" and second quantization (between corresponding states). That means we have to show

$$< k_i|A|k_j >=< 1_i| \sum_{lm} A_{lm} c_l^+ c_m|1_j > \tag{6.47}$$

which is immediately verified by Eq. (6.39). (The same argument holds for (6.45)).

The advantages of the occupation-number design become really powerful when dealing with excited states. A gas of N electrons in the ground state is given as

$$|\psi_0> = \underbrace{|1_1 \ldots 1_N}_{occupied} 0 \ldots > . \qquad (6.48)$$

Any excitations will show up by appearance of 1_ns at $n > N$. So, to monitor changes of the system, it suffices to write down only these parts

$$|\psi> = | \underbrace{\quad}_{omitted} n_{N+1} n_{N+2} \cdots > \qquad (6.49)$$

and the ground state can be considered a "vacuum" state now, called the "Fermi Vacuum":

$$|\psi> = | \underbrace{\quad}_{omitted} 00 >:= |0> . \qquad (6.50)$$

In this concept, particles are created out of nothing by applying c^+ onto the vacuum state (i.e. the system is excited). (At the same time, "holes" are created below the Fermi vector k_N).

The Green function in second quantization is obtained as follows: According to (6.27a), $G_0^+(k,k') = <k|G_0^+(t)|k'>$ is

$$G_0^+(k,k',t) = \frac{-i}{\hbar} <k|U(t)|k'> \Theta(t). \qquad (6.51)$$

We have, by virtue of Eqs. (6.39a,b), for the state in second quantization corresponding to $|k'>$, when $k' > k_N$

$$|k'> \,\hat{=}\, |1_{k'}> = c_{k'}^+|0> \qquad (6.52)$$

and

$$<k| \,\hat{=}\, <1_k| = <0|c_k = <0|U_0 U_0^{-1} c_k = e^{-i\omega_0 t} <0|U_0^+ c_k. \qquad (6.53)$$

In the last equality we have used the fact that U_0 is unitary

$$U_0(-t) = U_0^{-1}(t) = U_0^+(t). \qquad (6.54)$$

The energy of the ground state is $\hbar\omega_0$. Eq. (6.51) is now written

$$G_0^+(k,k',t) \hat{=} -\frac{i}{\hbar}e^{-i\omega_0 t} < 0|U^+(t)c_k U(t)c_{k'}^+|0 > \Theta(t) \qquad (6.55a)$$

$$G_0^+(k,k',t) \hat{=} \frac{i}{\hbar}e^{-i\omega_0 t} < 0| \quad c_k(t) \quad c_{k'}^+(0)|0 > \Theta(t) \qquad (6.55b)$$

where the transformation of operators from the Schrödinger to the Heisenberg picture $c(t) = U^+(t)cU(t)$ has been performed from (6.55a)to (6.55b). The latter lends itself to a plain interpretation: From the Fermi vacuum a "particle" (state) $|k' >$ is created at $t_0 = 0$, then at t a particle in state $|k >$ is destroyed. The probability amplitude for this process leading back to the vacuum is the Green function $G_0^+(k,k',t)$ times $\hbar$ (apart from a unimodular phase factor). Note that any single-particle basis can be used, e.g. $|r >$. Then the "states" are in fact particles at position r, and c and G_0 are now given as c_r and $G_0^+(r,r',t)$, and $< c_r^+c_r >$ is the standard particle density $\rho(r)$. (Probability of destroying and creating a particle at r'). Expression (6.55b) is termed "retarded" Green function G_0^+. This is because of the Heaviside step function in (6.55b) which guarantees that G_0^+ cannot propagate particles backwards in time as does G_0^-, the advanced Green function.

A slightly different interpretation, more common in textbooks on that subject, makes use of Eq. (6.51): Here, $G_0^+ \cdot \hbar$ is the modulus of the probability amplitude for finding an $(N+1)$-electron system in state $|1_k >$ at time t when it was in state $|1_{k'} >$ initially.

It should be kept in mind that the above is based upon the assumption of non-interacting free electrons. When a perturbation is introduced, such as an external field or interaction of electrons, the probability amplitude has to be built with the perturbed propagator, the series expansion (6.19) of which is still valid in many-electron systems.

From the physical interpretation given above a pictorial representation of G_0 can be contrived: The "propagating state" is represented by a trajectory in time, created at t_0 and destroyed at t (Fig. 6.2a). The limiting points are marked by states k, k' and times t, t_0 (Fig. 6.2b). In a more abstract way, the same can be imagined in (k,ω)-representation. Note that the trajectory has nothing to do with that of a *real* particle. The arrowed path just resembles a *process*. The corresponding probability amplitude cannot be inferred from the the graph. It is given by the explicit expression of G_0^+.

In particular, G_0^+ can be shown to vanish for $k \neq k'$ and $\omega \neq \omega'$ since momentum and energy are conserved. Consequently, we shall henceforth draw the graph for G_0 as in Fig. 6.2d and neglect processes with $k \neq k'$. This is no longer possible with the perturbed Green function as we shall see in the next section.

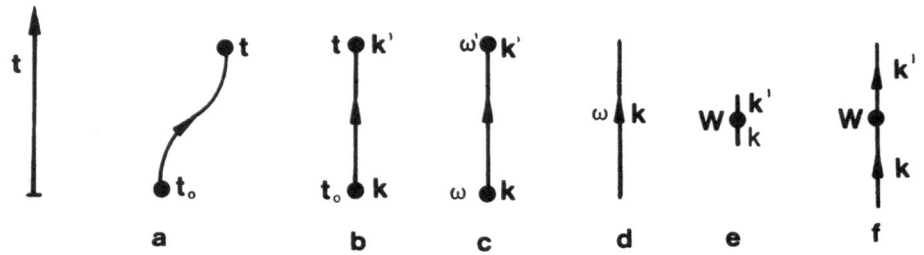

Fig. 6.2. Graphical representation of a "propagating" state. a) trajectory, b) symbol of process, c) same in ω-domain, d) restriction to $k' = k$, $\omega' = \omega$, e) scattering on W, f) connected graph.

6.9. The Perturbation Series in Graphical Representation

To calculate the perturbed propagator G—Eq. (6.19), one has to sum up the Born series

$$G = G_0 + G_0 W G_0 + G_0 W G_0 W G_0 + \dots \qquad (6.19b)$$

In (k, ω)-description, Eq. (6.19b) is a sum of products of matrices. For a given matrix element of G, each product contains a number of scalar terms from matrix multiplication. So, we have to add a huge number of elements, each of which we take the liberty to interpret as a "partial" probability amplitude. The expansion (6.19b) for an element of G, $G(k_m, k_l, \omega)$, say, is expressed in terms of matrix elements as

$$G(k_m, k_l, \omega) = G_0(k_m, k_l, \omega) +$$
$$+ \sum_{ij} G_0(k_m, k_i, \omega) \cdot W(k_i, k_j) G_0(k_j, k_l, \omega) + \ldots$$

$$(6.19c)$$

Now zoom your attention onto a particular term of the sum, i.e. fix your eyes on i, j for the moment. From the above, this term is the partial probability amplitude for the following process: a) propagation of state k_l to k_j, b) scattering from k_j to k_i on the potential W, and c) propagation from k_i to state k_m. The sequence of these three events corresponds to a product of "partial" probability amplitudes in (6.19c).—This is in complete analogy to the well-known statistical theorem for multiplication of probabilities of independent subsequent events, whereas the different possibilities to reach state k_m from k_l correspond to the sum over the probabilities to reach "intermediate" states k_i, k_j—again in analogy to the statistical theorem of adding probabilities.—In order to complete our list of symbols for processes, we introduce a graph for the scattering event on the perturbation W (Fig. 6.2e).

Obviously the number of terms proliferates enormously, when the series is expanded to higher and higher orders. One may easily go astray in the jungle of partial probability amplitudes—so in order to keep track of the calculation, it is desirable to use a simple though handy method of listing all possible contributions to the total probability. Instead, the *events* corresponding to partial probability amplitudes can be listed. Using the graphical representation for G_0 shown in Fig. 6.2d we have a direct resemblance between physics and mathematics since any connection of graphs such as in Fig. 6.2f means both a trajectory of a particle state and a multiplication of corresponding probability amplitudes.

We shall further elucidate this line of thought in the next section, by way of example.

116

6.10. An Example

For demonstration, we choose a static perturbation

$$W(k, k') = V_+ \delta(k - k' + q) + V_- \delta(k - k' - q) \qquad (6.56)$$

where $V_+ = V_-^*$. By Fourier transform, one can show that our example represents a real periodic potential with harmonic variation $\sin(qr)$.

As we have seen in the last section, a particular term of the series (6.19b) can be interpreted physically as a sequence of scattering processes of particles at either V_+ or V_-. Hence, the sum in (6.19c) reduces to two terms: $G_0(k + q, k + q, \omega) \cdot V_+ \cdot G_0(k, k, \omega)$ and $G_0(k - q, k - q, \omega) \cdot V_- \cdot G_0(k, k, \omega)$. Using the physical intuitive pictures Fig. 6.2d-f, this can be drawn as shown in Fig. 6.3

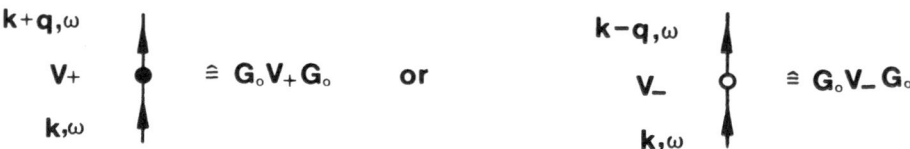

Fig. 6.3. Single scattering process with change of wave vector $+q$ or $-q$.

Due to the conservation of energy and momentum, G_0 does not change k or ω. Scattering on a static potential V_q changes k by $\pm q$. This is apparent in the drawing.

Observe that products in the Born series are represented by connecting propagators in line; the physical resemblance of this feature is the subsequent scattering of a particle at V_+, V_-.

Each path in the series stands for possible scattering processes of particles, albeit in (k, ω)-space, and the perturbed propagator G (which is likewise a probability amplitude), is given by the sum over contributions from all possible paths. Denoting $G(k, k', \omega)$ by ⇈, (6.19b) is drawn as

$$\text{⇈} = ↑ + ↑ + ↑ + ↑ + ↑ + ↑ + ↑ + \ldots \qquad (6.57)$$

It contains all possible ways a particle in state $|k>$ can evolve through time and be scattered by W.

Eq. (6.57) is much simplified when considering only a) diagonal elements of G and b) alternative scattering at V_+, V_-. b) is justified by the following "hand waving" argument: the particle remains as close as possible to its initial state when it is scattered alternatively on V_+, V_-. It can be surmised that these events contribute most to the diagonal element which we are interested in.

When the summation is restricted to those processes which lead back to k by alternative scattering to $k+q$, $k-q$ the series (6.57) is reduced to

$$G(k,k,\omega) \hat{=} \; \text{[diagram]} = \text{[diagram]} + \text{[diagram]} + \text{[diagram]} + \dots\dots\dots \qquad (6.58)$$

It is important to realize at this point that we can manipulate graphical expressions in a similar way as we can manipulate algebraic expressions.

This is pinpointed by the symbol "$\hat{=}$" which expresses that each term in the algebraic series of G has an equivalent on the right hand side. There are two essential operations on graphs, namely adding (denoted by a "+") and multiplication (symbolized by connecting two graphs). Upon this consideration, we can factor out $\omega \uparrow k \hat{=} G_0(k,\omega)$ from the sum, or we can express iterations of certain sequences in the path by powers. Hence,

$$G(k,k,\omega) \hat{=} \; \text{[diagram]} = \text{[diagram]} \; (1 + \text{[diagram]} + \left(\text{[diagram]}\right)^2 + \dots\dots\dots) \qquad (6.59)$$

which is an approximation for the perturbed propagator G. In diagrams,

$$G(k,k,\omega) \hat{=} \; \text{[diagram]} = \text{[diagram]} \; (1 + \sum_{n=1}^{\approx} \left(\text{[diagram]}\right)^n) = \frac{1}{k \text{[diagram]}^{-1} - \text{[diagram]}} . \qquad (6.60)$$

This is a graphical recapture of the Dyson Equation (6.6).

Explicitly, Eqs. (6.56) and (6.14) allow for a translation of Eq. (6.60):

$$G(k, k, \omega) \doteq \frac{1}{\hbar(\omega - \omega_k) - V_+ \frac{1}{(\omega - \omega_{k+q})\hbar} V_+^*}. \qquad (6.61)$$

The method of considering only parts of the propagator expansion, as exemplified above, is called "selective summation". It is among the most powerful ideas that have made the quantum mechanical description of many-body-effects feasible.

As discussed, the poles of G_0 constitute the resonant frequencies of the free electron system. The statement holds generally for any Green operator. A short calculation yields, for the present example, from Eq. (6.61)

$$\omega_k^\pm = \frac{\omega_k + \omega_{k+q}}{2} \pm \sqrt{\frac{(\omega_k - \omega_{k+q})^2}{4} + \frac{V_+ V_+^*}{\hbar^2}} \qquad (6.62)$$

where ω_k is the resonant frequency of the free electron in state $|k>$, see Eq. (6.2).

Eq. (6.62) describes the well-known splitting of an energy level in a perturbing field, such as a periodic ion-potential in a crystal.

For comparison of the perturbed frequency dispersion law with that of the free particle see Fig. 6.4.

An example is an electron in a periodic crystal potential. The appearance of an energy gap [3.7], [4.13] is well described; however, the selective summation (6.58) has introduced unrealistic behaviour for large and small $|k|$.

As we shall see in the next chapter, perturbation theory for the many-body system produces singularities in any order of the expansion. Only the selective summation can be considered a remedy because, under certain conditions, it is feasible to remove the singularities. However, there is no general criterion for whether or not the selected parts will converge. One has to learn about that by trial and error. In my opinion, this is evidence that physics is a game of chance—sometimes at least.

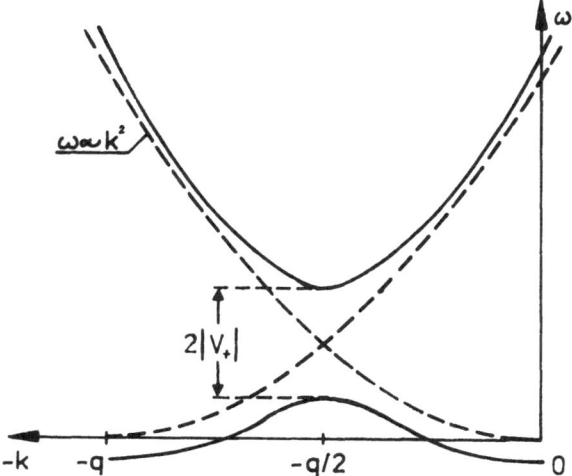

Fig. 6.4. "Perturbed" energy—momentum relation and the free electron energy parabola.

After these introductory remarks and the foregoing example, the electron gas will be investigated in the next chapter. Use is made of the diagrammatic technique, which reveals its full power when the perturbation is given by a two-particle operator, such as the Coulomb interaction, instead of the simple static potential just discussed.

6.11. The Coulomb Interaction

According to Eq. (6.45) the Coulomb interaction potential between two electrons in the Hamiltonian is written, in second quantization, as

$$W = \frac{1}{2} \sum_{q,k,l} V_{k+q,l-q,k,l} c^+_{k+q} c^+_{l-q} c_l c_k \qquad (6.63)$$

where the matrix element V_{ijkl} is

$$V_{ijkl} = < ij|V|kl > =$$
$$= 4\pi e^2 \iint d^3r d^3r' \varphi^*_i(r)\varphi^*_j(r') \frac{1}{|r-r'|} \varphi_k(r)\varphi_l(r'). \qquad (6.64)$$

Here we have introduced a convenient abbreviation: The subscripts i, j, k, l are short for k_i, k_j, k_k, k_l. We shall retain this notation for the remainder of the book, unless otherwise stated.—As already mentioned, the normalization volume is assumed to be unity.

Due to momentum conservation,

$$V_{ijkl} = <ij|V|kl> \delta_{i-k, l-j}. \tag{6.65}$$

The expectation value of one particular term of the sum in (6.63) in the vacuum state

$$V_{k+q, l-q, k, l} \quad < 0 | c_{k+q}^{+} c_{l-q}^{+} c_l c_k | 0 > /2$$

contains a process the physical resemblance of which is destruction of two particles in states k, l, whereas states $k+q, l-q$ are created by the interaction. This can be drawn

$$\cong \langle k+q \; l-q | V | k \, l \rangle. \tag{6.66}$$

The dashed line represents the Coulomb interaction by which momentum q was transferred.

The matrix element of the Coulomb interaction can be shown to be (e.g. [6.3], [6.4])

$$V_{k+q, l-q, k, l} = V_q = \frac{4\pi e^2}{q^2}. \tag{6.67}$$

Note that, according to the rules for drawing, the interaction consists of just two vertices "•" and the dashed "interaction" line. However the ends and beginnings of the particle propagators G_0 have been drawn, too, since they are necessary for calculation of the matrix elements.

Propagator series containing two-particle-interactions may now be drawn as (6.57), (6.58).

The diagrammatic description has been introduced by Feynman [6.3]. A complete theory has to take into account more details, such as advanced propagators G^-, time ordering, equivalence of particular diagrams. It should be mentioned that there is a difference between often used time-ordered ("Goldstone") diagrams and Feynman diagrams. We have left out all these details in our survey in favour of a crude understanding of the essential.

However, one thing is important to note in the next chapter: time increases from bottom to top in diagrams. Arrows showing top represent particles, arrows showing down represent holes, which are particles "propagating back in time".

A dictionary for translating Feynman diagrams is given at the end of the section.

One comment should be added to the table: The definition of $G_0(k, \omega)$ is slightly different from that previously given. The small quantity δ in the denominator shifts the poles of G_0 off the real ω-axis. This is equivalent to retaining the poles on the axis and defining $G_0^{+(-)}$ by particular path integrals as in Eq. (6.24). The advantage of the latter method is that there is only one Green operator in the ω-domain, whereas depending on the position of the poles above or below the real axis the former method distinguishes two G_0s. This is favourable in calculations since otherwise any integral over the real ω-axis (and there are many of them) demands a limiting procedure for the path integral.

Diagram element	Factor
$\mathbf{k}, \omega$ ⫫ or $\mathbf{k}, \omega$ ⫫	$iG(\mathbf{k}, \omega)$
$\mathbf{k}, \omega$ ↑ or $\mathbf{k}, \omega$ ↑	$iG_0(\mathbf{k}, \omega) = \dfrac{i}{\omega - \epsilon_k + i\delta_k}, \quad \begin{array}{l} \delta_{k>k_F} = +\delta \\ \delta_{k<k_F} = -\delta \end{array}$ with: $\displaystyle\int \frac{d\omega}{2\pi}[iG_0(\mathbf{k}, \omega)] = -\theta_{k_F - k}$ $\qquad\qquad\qquad\quad = -1, \quad k < k_F$ $\qquad\qquad\qquad\quad = \;\;\; 0, \quad k > k_F$
$\mathbf{k}$ $\mathbf{l}$ q, ϵ $\mathbf{m}$ $\mathbf{n}$	$-iV_{klmn}$ or $-iV_q$ (use $V_{klmn}(\epsilon)$ or $V_q(\epsilon)$ for time-dependent interaction)
Each fermion loop Example:	(-1)
Each intermediate energy parameter ω	$\displaystyle\int \frac{d\omega}{2\pi}$
Each intermediate momentum, $\mathbf{k}$	$\displaystyle\sum_k$ or $\displaystyle\int \frac{d^3k}{(2\pi)^3}$ (for $\Omega = 1$) (include sum over spins)

Table 6.1. Dictionary for translating Feynman diagrams. From [6.3].

7. Quantum Mechanical Description of the Electron Gas

In the following, the electron gas is described in terms of increasingly accurate approximations.

Seemingly, this is totally unrelated to what this monograph pretends to aim at. Still, there are two good reasons to go into these details: a) they are best suited to get acquainted with the methods of many-body theory introduced in the preceding chapter, and b) the discrepancies between results of simple model calculations of the ground state energy and experiment demonstrate quite directly why it is necessary to apply fairly sophisticated methods of calculation. So, Sections 7.1 to 7.5 should be considered as an introduction to what follows. Besides, they are useful as a repetition of fundamental properties of the electron gas which are discussed in more detail and accuracy in the literature on the subject—e.g. [7.1], [7.2], [7.3].

Sections 7.6 to 7.8 focus on the calculation of $\varepsilon(k, \omega)$ of the electron gas, thereby constituting the connection to the topics of this monograph.

We are mainly concerned with the ground state energy and the energy momentum relation. Throughout this chapter, the electron-gas will be treated in the jellium model, i.e. conduction electrons in a metal are considered, with the positively charged metal ions thought of as smeared out to a homogeneous neutralizing charge density.

7.1. The Jellium Hamiltonian

It is convenient to work with the dimensionless length r_s which is the average distance between electrons, measured in units of the Bohr radius $a_0 = \hbar^2/me^2 = 0.053nm$.

Obviously

$$(a_0 \cdot r_s)^3 \frac{4\pi}{3} = V_{e-} = \frac{1}{n} \tag{7.1}$$

in which V_{e^-} is the average volume occupied by an electron, and n is the electron density. For metals, $2 \leq r_s \leq 5.5$, corresponding to a Fermi energy between 12.5 and 1.7 eV.

The Hamiltonian, in second quantization, for an N-electron system is

$$H = \underbrace{\sum_i E_i c_i^+ c_i}_{free\ e^-\ H_0} +$$

$$+ \underbrace{\frac{1}{2} \sum_{ijk \neq 0} V_k c_{i+k}^+ c_{j-k}^+ c_j c_i + \frac{N(N-1)}{2} V_0}_{two-particle-interaction\ W} - \underbrace{\frac{N(N-1)}{2} V_0}_{jellium\ background} ,$$

$$(7.2a)$$

$$V_k = \frac{4\pi e^2}{k^2}. \qquad (7.2b)$$

V_k is the Fourier transform of the Coulomb potential. We have separated the term $k = 0$ in the two-particle interaction. It can be interpreted as the potential energy of N electrons in the constant part of the Coulomb field of $(N-1)$ other electrons. (Note that V_0 diverges!) The jellium background term emerges as follows: The potential energy of one electron in the field of a homogeneous positive charge density of total charge Ne is $-Ne^2 \int d^3 r/r = -NV_0$. Since there are a total of N electrons, the electron-background contribution is $-N^2 V_0$. In order to obtain the total background energy, we have to add N times the energy of one positive ion in the field of the $N-1$ remaining ions, which is $N(N-1)V_0/2$. The denominator stems from the demand that we must not count any ion-ion pair twice. Summing over both background contributions, we get approximately $(N \ll N^2)$ for the jellium background $-N(N-1)V_0/2$ which cancels the divergent term in the two-particle interaction W.

7.2. Sommerfeld Non-interacting e^--Gas

The most simple approximation completely neglects the two-particle interaction W. $H \approx H_0$. Hence, from (6.2) and (6.14)

$$G \approx G_0(k, \omega) = -\frac{1}{\hbar\omega - \frac{\hbar^2 k^2}{2m}}. \tag{7.3a}$$

The free particle dispersion is obtained from the poles of G_0

$$\varepsilon_i = \hbar\omega_i = \frac{\hbar^2 k_i^2}{2m}. \tag{7.3b}$$

The ground state energy/particle is

$$\bar{E} = \frac{2}{N} < H > \approx \frac{2}{N} < 0| \sum_i^{N/2} \varepsilon_i c_i^+ c_i |0 > =$$

$$= \frac{2}{N} \sum_i^{N/2} \varepsilon_i < 0| c_i^+ c_i |0 > = \frac{2}{N} \sum_i^{N/2} \varepsilon_i < n_i > = \tag{7.4}$$

$$= \frac{2}{N} \sum_i^{N/2} \varepsilon_i = \frac{2\hbar}{N} \sum_i^{N/2} \omega_i.$$

Here, the factor 2 comes from two particles with opposite spin/state. In k-representation each state occupies a volume of $(2\pi)^3/V$ (V is the volume in real space). Since there are $N/2$ states (spin $\uparrow, \downarrow$) we have

$$\frac{N}{2} \frac{(2\pi)^3}{V} = \frac{4\pi}{3} k_F^3. \tag{7.5}$$

For large N the sum (7.4) can be replaced by an integral, and from Eqs. (7.3b), (7.4), (7.5)

$$\bar{E} = \frac{2\hbar^2 V}{(2\pi)^3 N \cdot 2m} \int_0^{k_F} k^2 d^3 k = \frac{2\hbar^2}{N \cdot 2m} \int d\Omega \int_0^{k_F} k^2 dk k^2 =$$

$$= \frac{4\pi \hbar^2 k_F^5 2V}{2mN \cdot 5(2\pi)^3} = E_F \frac{k_F^3}{5n\pi^2} = \frac{3}{5} E_F. \tag{7.6}$$

From Eqs. (7.6), (7.5), (7.1) one obtains

$$\bar{E} = 29.8 r_s^{-2} \, eV \qquad (7.7)$$

which is, for metals, between approximately 1 eV and 8 eV.

The total energy of an excited state can be calculated similarly to $\bar{E}$ in Eq. (7.4). The excited state is written as

$$|\Delta \vec{k}> = |1_1 \ldots 0_{\vec{k}_0} 1_j \ldots 1_{\vec{k}_F} 0_l \ldots 1_{\vec{k}} 0_m \ldots > \qquad (7.8)$$

i.e. a particle at $k > k_F$ and a hole at $k_0 \leq k_F$ where $\Delta \vec{k} = \vec{k} - \vec{k}_0$. Note that subscripts $(j, l, m, k, \ldots)$ are vectors. Till the end of this section, the vector sign is explicitly written out on those states only which are involved in the following derivation.

Using the Hamiltonian H_0, the energy/particle of the excited state is calculated as was the ground state energy Eq. (7.4):

$$E(\Delta \vec{k}) = < \Delta \vec{k} | H | \Delta \vec{k} > = \frac{2}{N} (\sum_i E_i - E_{\vec{k}_0} + E_{\vec{k}}). \qquad (7.9)$$

The surplus energy over the ground state is

$$\Delta E(\Delta \vec{k}) := E(\Delta \vec{k}) - \bar{E} = \frac{\hbar^2}{2m} (k^2 - k_0^2). \qquad (7.10a)$$

For the wave vector difference $\Delta \vec{k}$, the inequality

$$|\Delta \vec{k} - \vec{k}_0| \leq k \leq \Delta \vec{k} + \vec{k}_0 \quad k_0 \leq k_F \qquad (7.10b)$$

holds, as can be seen from Fig. 7.1a. Obviously Eq. (7.10a) is not unambiguous when viewed as a function of wave *number* Δk. A range of energies belongs to a particular wave number Δk.

For given Δk, the energy ΔE is bounded by the possible boundary values of k, which are (see Fig. 7.1a)

$$\Delta E + E_F \leq \frac{\hbar^2}{2m} ((\Delta k + k_0)^2 - k_0^2 + k_F^2) \leq \frac{\hbar^2}{2m} (\Delta k + k_F)^2, \qquad (7.11a)$$

$$\Delta E + E_F \geq \frac{\hbar^2}{2m} ((\Delta k - k_0)^2 - k_0^2 + k_F^2) \geq \frac{\hbar^2}{2m} (\Delta k - k_F)^2. \qquad (7.11b)$$

Hence

$$\frac{\hbar^2}{2m} (\Delta k - k_F)^2 \leq \Delta E + E_F \leq \frac{\hbar^2}{2m} (\Delta k + k_F)^2 \qquad (7.12)$$

from which the range of possible excitation energies ΔE and wave number transfers Δk in the gas is determined (see Fig. 7.1b).

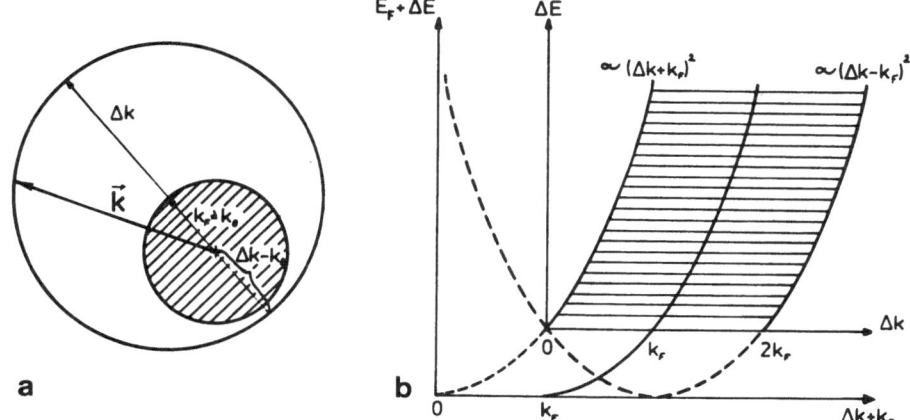

Fig. 7.1. a) Fermi sphere (hatched) and possible final wave vector, $\vec{k}$ for given Δk and $\vec{k}_0 = \vec{k}_F$. b) possible excitation energies ΔE as a function of Δk (hatched). The hatched area is called electron-hole continuum.

7.3. Hartree Approximation (HA)

A better approximation consists of taking into account only the constant background ($q = 0$) of the interaction in Eq. (7.2):

$$V_0 = \sum_{ij} < ij|V|ij > c_i^+ c_j^+ c_j c_i, \qquad (7.13a)$$

in diagram form this is:

$$V_0 \,\hat{=}\, \qquad (7.13b)$$

V_0 acts as a scattering potential like that in our static example in the foregoing chapter. The difference is that V_0 is a *two-particle* potential; thus we have two incoming and two outgoing particle lines, whereas what we need in the expansion of the single particle propagator is a graph with one incoming and one outgoing line in order that the particle in question can be considered as scattered off a potential. Let particle i be our test particle. For i, the scattering takes place at the left vertex in the diagram. What happens to particle j is of no relevance for the test particle. We may trace

j's life and find some more interactions in its world-line. The only important thing to notice is that the propagator for j has to vanish eventually by recombination with some hole line—otherwise we would not be allowed to speak of a *one-particle* propagator for i. The most simple way to do so is to close the j-line without any interaction:

$$(7.13c)$$

Note, by the way, that j is a hole. It can be shown that terms where j is a *particle* must vanish [6.4]. The closed hole line is called a "bubble".

The physical interpretation of this diagram is that a particle enters in state i, knocks another particle out of state j (thus creating a hole) and knocks it back into j at the same instant. The test particle emerges from the process in the same state as it entered, i.e. it has neither changed energy nor momentum. Therefore such a process is called *forward scattering*. At the first glance it looks as if nothing has happened to the test particle. However, from the fact that neither energy nor momentum of the particle has changed we may not infer that nothing has happened. Remember that the probability amplitudes do not show up in the graph, which will influence the propagator expansion in a definite way. The particle remains in the same state but the probability for this to be the case may differ from unity.

According to the rules for translating diagrams into expressions, Table 6.1, the diagram (7.13c) is equivalent to

$$[iG_0(i,\omega)]^2 \cdot \sum_{j<k_F} (-iV_0) \cdot (-1) \cdot (-1) \qquad (7.13d)$$

where the first phase factor (-1) stems from the integration over the bubble and the second one from the closed fermion loop.

When only V_0 is inserted in the expansion of the propagator, the following series is obtained

$$\| = \uparrow + \uparrow\!\!-\!\!-\!\!\bigcirc + \updownarrow\!\!-\!\!-\!\!\begin{matrix}\bigcirc\\\bigcirc\end{matrix} + \updownarrow\!\!-\!\!-\!\!\begin{matrix}\bigcirc\\\bigcirc\\\bigcirc\end{matrix} + \qquad (7.14)$$

where the hole lines on the right had to be closed to a bubble since these virtually created holes are destroyed by the same interaction due to conservation of particle number. Eq. (7.14) diverges since $V_0 \propto \lim_{q \to 0} 4\pi e^2/q^2$. However, the Hartree-term is compensated by the jellium background, as we have argued at the beginning of this chapter after Eq. (7.2). Consequently, the perturbed propagator equals the free electron propagator in Hartree approximation. There is no difference to the Sommerfeld free electon approximation, as to what concerns the ground state energy.

The connection to the Hartree self consistent field (SCF) approach is established as follows: $\vdash\!\!-\!\!-\!\!\bigcirc$ plays the same role as $\bullet\!\!\uparrow$ in (6.52) so it can be considered a scattering process in an "effective" single particle potential $\bullet\!\!\uparrow \hat{=} <k|V_{eff}|k>$:

$$\begin{matrix} k \\ \bullet\!\!-\!\!-\!\!-\!\!\bigcirc \\ k \end{matrix} \hat{=} \begin{matrix} k \\ \bullet\!\!\uparrow \\ k \end{matrix} \hat{=} \langle k|V_{eff}|k\rangle \qquad (7.15)$$

In r-representation, Eq. (7.15) is (cf. 6.57):

$$\int d^3 r \varphi_k^*(r)\varphi_k(r) \sum_j \int d^3 r' \varphi_j^*(r')\varphi_j(r')V(r-r') \hat{=}$$

$$\hat{=} \int d^3 r \varphi_k^*(r)\varphi_k(r)V_{eff}(r). \qquad (7.16)$$

Hence,

$$V_{eff}(r) = \sum_j \int d^3 r' \varphi_j^*(r')\varphi_j(r')V(r-r'). \qquad (7.17)$$

In the present approximation, the Schrödinger Eq. reads

$$\left(\frac{\hbar^2}{2m}\nabla^2 + V_{eff}(r)\right)\varphi_k(r) = E_k\varphi_k(r). \qquad (7.18)$$

The pair of equations (7.17), (7.18)—the Hartree equations—can be solved iteratively, starting with the free electron wave function for φ_k until φ_k is "self consistent". This is the Hartree-SCF formalism.

7.4. Hartree-Fock Approximation (HFA)

Whereas in HA, Eq. (7.13), the sequence of particles i, j in the bras and kets was identical in any term of the sum, now we allow for their exchange:

$$W = \frac{1}{2} \sum_{ij} < ij|V|ij > c_i^+ c_j^+ c_j c_i + < ij|V|ji > c_i^+ c_j^+ c_i c_j. \quad (7.19)$$

Note that, contrary to HA, Fourier coefficients $V_q = V_{i-j}$, $q \neq 0$ are involved in the rightermost part (cf. 6.60). In diagrams, W is given as

$$(7.20)$$

and the propagator series is

$$(7.21)$$

The frequency dispersion law is inferred from the poles of Eq. (7.21):

$$\hbar\omega = \hbar\omega_i - \sum_{j<k_F} (\underbrace{< ij|V|ij >}_{V_0} - \underbrace{< ij|V|ji >}_{\frac{4\pi e^2}{|i-j|^2}}). \quad (7.22)$$

Evaluation of Eq. (7.22) yields, replacing the state descriptor i with the corresponding wave number k [6.4]

$$\omega(k) = \frac{\hbar k^2}{2m} - \frac{e^2 k_F}{2\pi}(2 + \frac{k_F^2 - k^2}{k k_F} ln\frac{|k + k_F|}{|k - k_F|}). \quad (7.23)$$

From this, the effective mass at the Fermi-wave number

$$m^* := \hbar k_F / \left(\frac{\partial \omega}{\partial k} \right)_{k_F} \tag{7.24}$$

tends to 0 in HFA, since $\lim_{k \to k_F} \partial \omega / \partial k \to \infty$, which certainly does not correspond to reality. HFA does not seem to be so good an approximation.

Another interesting property for testing the accuracy of a model is the width of the conduction band in metals. For non-interacting electrons and in HA this is simply the Fermi energy. The interaction changes E_F to $E_0 = \hbar \omega(k_F)$, see Eq. (7.23).

As displayed in Table 7.1, HFA is worse than HA although it is the more precise approximation.

The ground state energy/particle is given in first order perturbation theory as

$$\bar{E} = \frac{1}{N} < 0|H_0|0 > + \frac{1}{N} < 0|W|0 > \tag{7.25}$$

with W from Eq. (7.19). The first term ("zeroth order") has already been calculated in Eq. (7.6). The second yields $-12.4/r_s eV$ [6.5], hence

$$\bar{E} \approx (29.8 r_s^{-2} - 12.4 r_s^{-1})eV. \tag{7.26}$$

$\bar{E}$ as a function of r_s is drawn in Fig. 7.2.

The minimum implies that the system is bound, contrary to the HA, which is an important improvement. The experimental fit is rather good ($r_s = 3.96, \bar{E} = -1.13eV$ for Na), but far from satisfactory.

Any further improvement over Eq. (7.26) is called correlation energy, due to the fact that correlations in the movements of individual particles should be included to improve the result. The usual way is to proceed to higher orders in perturbation theory.

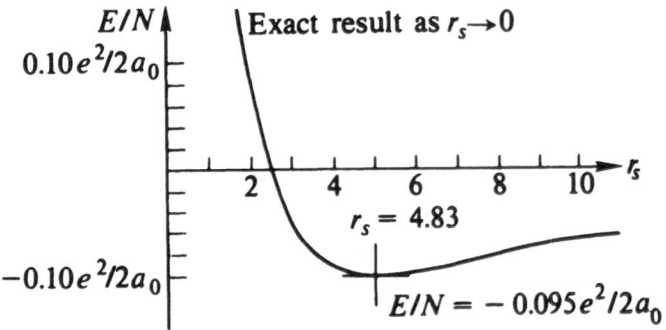

Fig. 7.2. HFA-ground state energy of an electron gas in the jellium model. From [7.1].

7.5. Why Higher Order Perturbation Theory Does not Work

When one proceeds to higher order in the Coulomb interaction, a new type of diagram will occur. For instance, the second order term in the perturbation expansion of the ground state energy contains products of matrix elements $< 0|W|n >< n|W|0 >$ in which $|0 >$ is the Fermi vacuum and $|n >$ is any excited state. As we have seen, there is a one-to-one correspondence between a particular process probability amplitude and a matrix element. The basic event involved in the aforementioned product consists of creating two particle-hole pairs from the Fermi vacuum and subsequently destroying it, by the Coulomb interaction. This can be drawn as

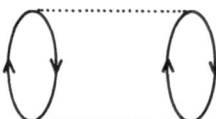

whereas the first order term contained only *one* Coulomb line:

Similar "second order" graphs will appear in the expansion of the perturbed propagator:

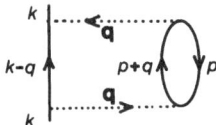

which is to say: the test particle in state k creates, by transfer of momentum q via the Coulomb interaction, a particle-hole pair. The pair is destroyed subsequently. Graphs of this type turn out to diverge since they contain $1/q^2$ for each interaction line. Translation via the dictionary Table 6.1 yields

$$-1 \sum_p \iint d\omega' d\omega'' \left(\frac{4\pi e^2}{2\pi}\right)^2 iG_0(k-q, \omega-\omega')$$

$$iG_0(q+p, \omega'+\omega'') \cdot \qquad (7.27)$$

$$iG_0(p, \omega'') \sum_q \frac{-i}{q^2}\frac{-i}{q^2} \to \int_0^{k_F} \frac{d^3q}{q^4} \to \infty$$

Divergence becomes even worse for the higher order diagrams. The most diverging parts are those with *equal* momentum transfer in each interaction since denominators factorize.

Since $V_q \propto 1/q^2$, there are terms $\propto (1/q^2)^n$ in n-th order perturbation theory. A detailed consideration [7.2] shows that these terms create singularities of ever increasing order. The ground state energy diverges in any order of perturbation theory (Of course, in infinite order, the series converges,— think of an alternating scalar series for analogy).

To summarize our results obtained thus far:
- The HFA explains why the jellium model resembles a bound system —insofar HFA is physically reasonable.
- in HFA, the effective mass at the Fermi edge is infinite, an unrealistic result.
- HFA cannot explain the width of the conduction band of metals.
- any attempt to improve the ground state energy by a perturbation expansion in higher powers of the Coulomb potential fails. This breakdown of a celebrated formalism has set back the systematic investigation of the electron gas for some two

134

decades. It was therefore justified to look for a more appropriate approach in order to overcome the difficulties with the band width, the effective mass and the ground state energy.

About 1950 D. Bohm and D. Pines approached the problem by a canonical transform of the Hamiltonian [7.4]. Since the divergence in the perturbation series arises from the long range interaction of the Coulomb potential $(q \rightarrow 0)$ they separated this very part in the Hamiltonian. A canonical transformation then delivers eigenstates of the long range part of V. What remains is a screened Coulomb potential

$$H = H_0 + (\sum_{ij,k<k_L} + \sum_{ij,k \geq k_L}) < i + k, j - k| \frac{v_k}{2} |ij> . \qquad (7.28)$$

The second sum describes a short range interaction via a screened Coulomb potential V_{eff}, and the first sum over $k < k_L$ contains the long range interaction the eigenstates of which have been obtained by a canonical transformation by Bohm and Pines [7.4].

The splitting into 2 parts in r-representation is shown in Fig. 7.3.

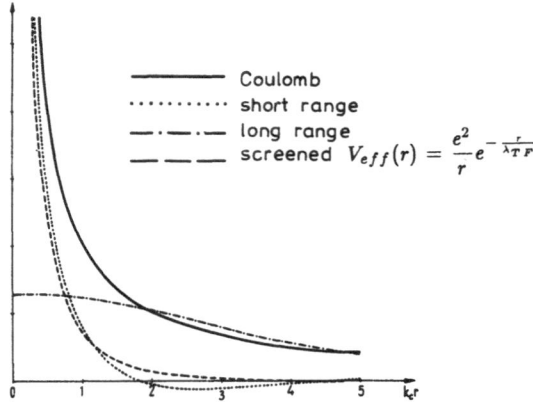

Fig. 7.3. Long- and short-range parts of the Coulomb potential. From [7.3].

A powerful method to deal with the difficulties mentioned is the diagrammatic technique. We have already seen that by selective summation, approximations such as HA or HFA can be interpreted simply. So instead of using the Bohm-Pines approach we shall present a more intuitive picture using Feynman diagrams. For reasons which will soon become clear this approximation is called random phase approximation.

7.6. The Random Phase Approximation (RPA)

Whereas in the HA only $\vdash\!\!\cdots\!\!\bigcirc$ and in the HFA $\vdash\!\!\cdots\!\!\bigcirc\!+\!\backsim\!\!\smile$ were used in the propagator expansion (6.28), the inclusion of all interactions would yield G exactly: we could draw the series as

$$\|\ =\ \uparrow\ +\ \Sigma\!\!\!\bigcirc\ +\ \begin{array}{c}\uparrow\\\Sigma\!\!\!\bigcirc\\\uparrow\\\Sigma\!\!\!\bigcirc\\\uparrow\end{array}\ +\ \cdots\cdots \tag{7.29}$$

Taken for granted that the series converges, we have

$$\|\ =\ \frac{1}{\uparrow^{-1}-\Sigma^{*}\!\!\!\bigcirc} \tag{7.30}$$

which is (Eq. 6.16) in diagram form. Since the zeros of the denominator of Eq. (7.30) determine the system's energy, we see that the formerly free particle is now "dressed" with $\sum^{*}$, so to speak. From (7.29) it is formally clear that $\sum^{*}$ is the sum total over all diagrams with two-particle interaction which are *not* repetitions of more simple diagrams. For instance, $\vdash\!\!\cdots\!\!\bigcirc$ belongs to $\sum^{*}$, but $\vdash\!\!\cdots\!\!\overset{\circ}{\circ}$ does not (it is a part of $\sum^{*2}$). If one thinks it over twice it becomes evident that

$$\Sigma^{*}\!\!\!\bigcirc\ =\ \vdash\!\!\cdots\!\!\bigcirc\ +\ \backsim\!\!\smile\ +\ \boxed{}\bigcirc\ +\ \bigtimes\ +\ \bigg|\ \overset{\circ}{\circ}\!\!\bigcirc\ +\ \bigg|\ \bigcirc\ +\ \cdots\cdots \tag{7.31}$$

$$\underbrace{\qquad\qquad\qquad\qquad\qquad\qquad\qquad\qquad\qquad}_{\text{irreducible self-energy diagrams}}$$

$\sum^{*}$ is called proper self-energy.

136

From Eq. (7.31) the divergence problem encountered previously becomes apparent: Since each Coulomb interaction (······) resembles a factor $\propto 1/q^2$ we find that any finite approximant for $\sum^*$ diverges; therefore, a perturbation expansion in any power of $1/q^2$ for the proper self-energy (and consequently for the ground-state energy) fails. We may, however, join the game of hazard, which is called selective summation, invented by R. Feynman. The aim is to obtain a convergent series—which, moreover, is to be a good approximation for $\sum^*$— by selecting certain diagrams out of Eq. (7.31). The rule of the game is to select the *most* divergent diagrams (yes—the most!) which are exactly those where denominators factorize:

As is often the case, fortune favours the brave—our tour de force is successful. We obtain

$$(7.32a)$$

which is just the random phase approximation (RPA) so called because no correlation between the phases of virtual particle-hole pairs exists — (only one pair is created at one instant). The diagrams constituting (7.32) are called ring diagrams.

After factoring out the unperturbed propagator ↑ and the Coulomb interaction ········ , the sum is

$$(7.32b)$$

The rightermost term is called effective interaction since it replaces all the virtual processes between the first and the last q transfer by a simple (as if) interaction. It is designated

$$\underset{\text{\tiny RPA}}{:::::::::::::} := \frac{\cdots\cdots\cdots}{1- \cdots 0} \qquad (7.33a)$$

or, translated

$$V_{eff,RPA}(q,\omega) = \frac{V_q}{1 + V_q \pi_0(q,\omega)} \qquad (7.33b)$$

where V_q is the Coulomb potential (6.67) and π_0 is the free electron polarizability (see table 6.1 for translation)

$$\pi_0(q,\omega) = -\frac{2}{(2\pi)^4} \int d^3 k d\omega' \frac{i}{(\omega + \omega' - \omega_{k+q} + i\delta)} \frac{i}{(\omega' - \omega_k - i\delta)}. \qquad (7.34)$$

Physically, (7.33b) means that the Coulomb potential V_q of a free particle is replaced by V_{eff} in the interacting electron gas, i.e. an interacting particle in the gas behaves as if it were free, but exerting V_{eff}!

Observe that the gradients of $V_q, (V_{eff})$ are the displacement field $\vec{D}$ and the electric field $\vec{E}$, respectively in the nonretarded (longitudinal) case (see Sect. 4.3). Hence, taking the gradients in Eq. (7.33b),

$$\vec{E}(q,\omega) = \frac{\vec{D}(q,\omega)}{1 + V_q \pi_0(q,\omega)}. \qquad (7.33c)$$

By comparison with Eq. (4.12a),

$$\vec{D} = \tilde{\varepsilon} * \vec{E} \leftrightarrow \vec{D}(\vec{k},\omega) = \varepsilon(\vec{k},\omega)\vec{E}(\vec{k},\omega) \qquad (4.12a)$$

the denominator of (7.33c) can be identified with a generalized dielectric function ε of the medium

$$\varepsilon(q,\omega)_{RPA} = 1 + V_q \pi_0(q,\omega). \qquad (7.35)$$

The polarizability π_0, Eq. (7.34) can be evaluated analytically.

In the static limit $\omega = 0$, [2.4]

$$V_{eff} = \frac{4\pi e^2}{q^2 + \frac{1}{\lambda^2}} \qquad (7.36a)$$

with

$$\frac{1}{\lambda^2} = \frac{1}{\lambda_{TF}^2} \cdot \left[\frac{1}{2} - \frac{2}{x}(1 - \frac{1}{4}x^2) \ln \left| \frac{2-x}{2+x} \right| \right], \qquad x = q/k_F \quad (7.36b)$$

which is, for $q \ll k_F$,

$$V_{eff}(q \ll k_F, \omega = 0) = \frac{4\pi e^2}{q^2 + \frac{1}{\lambda_{TF}^2}}. \qquad (7.37)$$

The Fourier transform of (7.37) is a screened potential V_{eff} displayed in Fig. 7.3. The Fourier transform of the exact static potential (7.36) shows an oscillation superimposed onto the exponential decay. The oscillation stems from the singularity at $q = 2k_F$ in the derivative of $1/\lambda^2$. It can be shown [2.4], [7.1] that the oscillations are proportional to $\cos(2k_F r)/r^3$ as $r \to \infty$. They have been predicted by Friedel [5.2]. Via Poissons equation they are connected to oscillations of the charge density and may therefore, under certain circumstances, be observed by electron scattering experiments [5.4].

$1/\lambda_{TF}$ is termed Thomas-Fermi screening length, and is the quantum analogy to the Debye length in classical plasmas. Electrons interacting via V_{eff} can be viewed as "quasi-electrons".

The first calculation of $\varepsilon(q, \omega)$ from (7.34, 35) was made by Lindhard [7.5]. The formulae are rather complicated and not given here. The qualitative behaviour of $Re(\varepsilon)$ and $Im(\varepsilon)$ is depicted in Fig. 7.4. For $0 < q \ll k_F$ (as assumed in the figure) there is a linear increase in ε_2, starting at $\omega = 0$, then followed by a quadratic decrease to zero. In this frequency domain, excitation of particle-hole pairs is possible—leading to dissipation of energy. In the same domain, ε_1 deviates from the classical Drude-behaviour Eq. (4.34a), whereas the classical high energy increase is not much influenced.

Along the path $\omega_c(k)$, which is in detail

$$\omega_c(k) = \omega_p(1 + \frac{3v_F^2}{10\omega_{pl}^2}k^2 + \ldots) , \qquad (7.38)$$

$$\omega_p^2 = \frac{4\pi n e^2}{m}, \qquad (7.39)$$

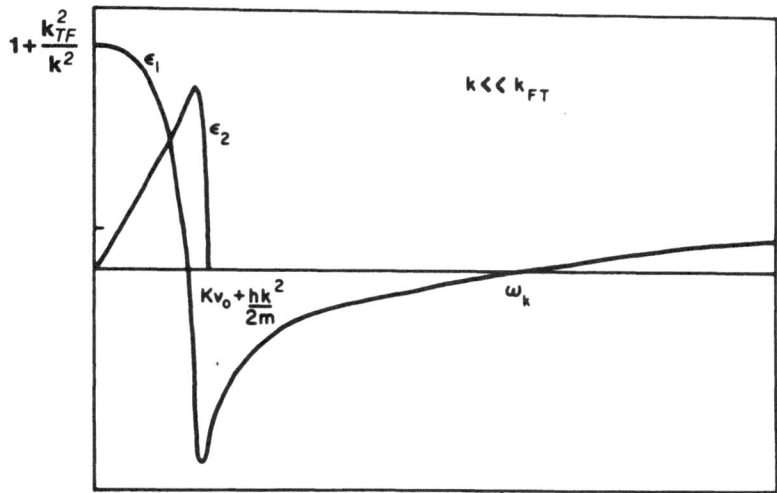

Fig. 7.4. The general behaviour of ε_1 and ε_2 calculated in the RPA for the free-electron gas. From [7.2].

ε vanishes. According to Eq. (7.33b), V_{eff} in RPA diverges at ω_c, i.e. (7.38), (7.39) define a resonant frequency. The resonance is caused by the long-range part of the Coulomb interaction, which establishes collective oscillations (plasmons) at $\omega_c(k)$. This result has already been derived in the framework of Maxwell theory (Chapter 4) by the same line of thought.

It is perhaps disappointing that eventually, after bothering so much with Feynman diagrams, we resort to the classical argument $\varepsilon = 0$ so as to find collective oscillations. This would have been inferred without diagrammatic magic, too. However, important facts have been worked out which could not have been obtained classically. First note that neither the screening length nor the accurate functional form of ε can be obtained classically (especially the q-dependence). Moreover, one may proceed to calculate the proper self energy $\sum_{RPA}^{*}$ from (7.32), and the frequency dispersion law from the poles of (7.30).

One arrives at [6.4]

$$\omega_k = \frac{\hbar k^2}{2m} - 0.166 r_s (\ln r_s + 0.203) \frac{\hbar k k_F}{2m} + const. \qquad (7.40)$$

which obeys, near k_F, a free-particle like behaviour

$$\omega_k \approx \frac{\hbar k^2}{2m^*} \quad with \quad m^* = \frac{m}{1 - 0.008 r_s (\ln r_s + 0.203)} \qquad (7.41)$$

thus overcoming the HFA breakdown apparent in Eqs. (7.23), (7.24).

Table 7.1 compares the band widths of some metals with calculations obtained from the approximations discussed previously. Not only does RPA suppress the HFA singularity at the Fermi surface (7.23), (7.24) but it also yields fairly good agreement of E_0 with experiment.

r_s	$E_F(eV)$	Exper.	HA	HFA	RPA	
2.49	4.8	3.7	4.8	8.5	2.7	Li
1.43	14.6	13.8	14.6	20.7	12.7	Be
1.58	11.9	11.8	11.9	20.0	9.8	Al

Table 7.1. Width E_0 of conduction bands in some metals.

Although experimental results do not exactly compare with the RPA prediction, it is at present one of the best understood, highly successful and commonly used approaches to the description of the interacting electron gas.—An approximation beyond RPA will be discussed in the next chapter.

7.7. Electron Scattering

In Chapter 5, we have derived, by classical arguments, a relation between the scattering cross section and the dielectric function ε Eq. (5.45). This relation can be shown to remain valid quantum-mechanically [2.4]. Consequently, calculation of ε allows for a prediction of electron scattering cross sections by means of Eq. (5.45): The cross section reveals poles at the zeros of ε. In particular, scattering on the interacting electron gas is described approximately by Lindhard's calculations (cf. Fig. 7.4). Scattering takes place a) in the electron-hole continuum by a screened interaction and b) along the plasmon line in the (k, ω)-plane given by Eq. (7.38) as we have already mentioned.

There is another more transparent approach to the prediction of scattering cross sections which shall be given in the following.

As we have seen in Chapter 2, the inelastic scattering cross section is given in terms of the dynamical structure factor as

$$\frac{\partial^2 \sigma}{\partial E \partial \Omega} = \left(\frac{2me^2}{\hbar^2 q^2} \right)^2 \frac{k_b}{k_a} \cdot |S(\vec{q}, \omega)|^2. \tag{2.14}$$

The latter is, due to Eq. (2.18), the Fourier transform of the electron density-density product.

$$|S(\vec{q}, \omega)|^2 = \int dt e^{i\omega t} \int d^3 r \int d^3 r' e^{-i\vec{q}(\vec{r} - \vec{r}')} < n(\vec{r}, t) n(\vec{r}', 0) > . \tag{7.42}$$

The expectation value on the right hand side can be written alternatively as

$$< 0|c_{\vec{r}}^+(t) c_{\vec{r}}(t) c_{\vec{r}'}^+(0) c_{\vec{r}'}(0)|0 > = G_2(\vec{r}, t, \vec{r}, t, \vec{r}', 0, \vec{r}', 0), \tag{7.43}$$

which is a particular case of the two-particle Green function G_2. In complete analogy to the interpretation of the one-particle Green function (the "usual" propagator) Eq. (6.55), the two-particle Green function G_2 is the probability amplitude that a particle—hole *pair* created at $\vec{r}', t_0 = 0$ be propagated to $\vec{r}, t$.

In diagrams, Eq. (7.43) consists of all possible ways of creating a particle-hole pair at $\vec{r}', t' = 0$ and subsequently destroying it at $\vec{r}, t$:

$$\text{a} \quad \text{b} \quad \text{c} \quad \text{d} \quad \text{e} \quad \text{f} \qquad \hat{=} \ G_2 \quad (7.44)$$

From Eqs. (2.14), (7.42) and (7.43), it is obvious that the poles of $\partial^2\sigma/\partial E\partial\Omega$ coincide with poles of G_2 in (k,ω) -representation. Such poles are related to charge-density fluctuations by Eq. (7.42): Any periodic variation of $n(\vec{r})$ will cause $|S|^2$ to diverge at a particular wave vector. Hence, it makes no difference to state that the scattering probability is determined by the poles of the two-particle Green function G_2 or by resonant excitation of charge density waves by the probe electron.

7.8. Polarization Diagrams

Obviously there is still another way of expressing this fact since G_2 is a polarization diagram, as can be seen by analogy to π_0, Eq. (7.34). We may express G_2 by a sum over products of *proper* polarization diagrams $\textbf{\textit{O}}^*$ which are *not* repetitions of more simple diagrams (i.e. which cannot be split into separate parts by *one* cut through an interaction line (·········). See also the argumentation preceding Eq. (7.29). Now,

$$(7.45)$$

where terms a), b), d), e) of Eq. (7.44) belong to $\textbf{\textit{O}}^*$, but not so c) or f)!

The infinite series (7.45) yields

$$(7.46)$$

Taking only the first term of the proper polarization, viz. the free electron polarizability $\bigcirc = \pi_0$, we have

$$\bigcirc = \frac{\bigcirc}{1 - \cdots\bigcirc} \quad , \quad G_2 \approx \frac{\pi_0}{\varepsilon_{RPA}} \tag{7.47}$$

The denominator has already been identified with the dielectric permittivity (cf. Eqs. (7.33), (7.35)). Eq. (7.47) relates the two-particle Green function G_2 (and hence the scattering probability) with the dielectric function.

In the end, we have shown that it makes no difference whatsoever if we look for the poles of the scattering cross section, the two-particle Green function, the polarization or the zeros of the dielectric function.

8. Beyond Simple Models

The closing chapter is dedicated to a discussion of some details not yet covered. We shall outline how inhomogeneities on an atomic scale in a medium, influence the dielectric function, how ion-electron interactions can be included into the random phase approximation, thus surpassing the simple jellium model described in the foregoing chapter, and what the consequences of multiple inelastic scattering of electrons are. These issues ought to be considered appendices to the mainframe of the present monograph. By no means do they represent a completion of important facts on electron interactions, or even a list of the most important extension of simple models.

Instead they are intended to give an impression of which lines of thought may lead from the preceding chapters to genuine work in this field. The selection of examples is based on the author's experience with particular problems and his personal insight difficulties. The reader should not infer that there *are* any problems contained in the choice given—any of the sections can be skipped without having a bad conscience.

8.1. Solid State Effects

In Chapter 4 it was shown that the dielectric displacement is given in linear response theory, and for isotropic, homogeneous systems, by

$$\vec{D} = \tilde{\varepsilon} * \vec{E} \longleftrightarrow \vec{D}(\vec{k},\omega) = \varepsilon(\vec{k},\omega)\vec{E}(\vec{k},\omega), \qquad (4.12a)$$

i.e. perturbations of different wave vector or frequency do not couple with one another. A monochromatic external perturbation thus induces a polarization of the same periodicity in space and time.

In what follows, we shall derive a similar relation for lattice periodic systems. It will turn out that external perturbations can induce fields of higher frequency in space since Fourier coefficients $\vec{D}(\vec{k}), \vec{E}(\vec{k'})$ with different wave vectors are then coupled.

For simplicity, we resort to isotropic crystals: longitudinal and transverse components of $\vec{D}(\vec{r}), \vec{E}(\vec{r})$ are still decoupled. So, $\tilde{\varepsilon}$ can be considered a scalar quantity, and D, E are either longitudinal or transverse components of the fields.

Again we start with Eq. (4.10), but $\tilde{\varepsilon}$ no longer contains *differences* of space vectors, due to the inhomogeneities. Rather we have

$$\tilde{\varepsilon} = \tilde{\varepsilon}(\vec{r}, \vec{r'}, t - t'), \tag{8.1}$$

and by passing over from (4.10) to Fourier transforms:

$$D(\vec{k}, \omega) = \frac{1}{(2\pi)^3} \int d^3k' \varepsilon(\vec{k}, \vec{k'}, \omega) E(-\vec{k'}, \omega), \tag{8.2}$$

where the double Fourier transform of the dielectric function

$$\varepsilon(\vec{k}, \vec{k'}, \omega) = \int d^3r \int d^3r' \int dt \tilde{\varepsilon}(\vec{r}, \vec{r'}, t) e^{-i(\vec{k}\vec{r} + \vec{k'}\vec{r'} - \omega t)}. \tag{8.3}$$

has been used. For translational invariance:

$$\varepsilon(\vec{k}, -\vec{k'}, \omega) = (2\pi)^3 \delta^3(\vec{k} - \vec{k'}) \bar{\varepsilon}(\vec{k'}, \omega) \tag{8.4}$$

formula (4.12) is regained from Eq. (8.2).

In the case of lattice periodicity, ε can be written alternatively as $\varepsilon(\vec{r}, \vec{r} - \vec{r'}, t)$, which is lattice periodic with respect to the first variable. This can be visualized by keeping $\vec{r} - \vec{r'}$ fixed and varying $\vec{r}$. Hence, one may expand $\tilde{\varepsilon}$ into a Fourier series with respect to $\vec{r}$:

$$\tilde{\varepsilon} = \sum_j \tilde{\varepsilon}_j(\vec{r} - \vec{r'}, t) e^{i\vec{G}_j \vec{r}}, \tag{8.5}$$

$\vec{G}_j$ are reciprocal lattice vectors.

Substituting (8.5) into (8.3) yields

$$\varepsilon(\vec{k}, -\vec{k'}, \omega) = (2\pi)^3 \sum_i \delta^3(\vec{k} - \vec{k'} - \vec{G}_i) \varepsilon_i(\vec{k'}, \omega), \tag{8.6}$$

(ε_i is, in our notation, the Fourier transform of $\tilde{\varepsilon}_i(\vec{x}, t)$), and (8.2) is

$$D(\vec{k}, \omega) = \sum_i \varepsilon_i(\vec{k} - \vec{G}_i, \omega) E(\vec{k} - \vec{G}_i, \omega). \qquad (8.7)$$

We may write $\vec{k} = \vec{k}_0 + \vec{G}_j$ with $\vec{k}_0$ restricted to the first Brillouin zone. Any function $f(\vec{k}) = f(\vec{k}_0 + \vec{G}_j)$ can now be designated $f_j(\vec{k}_0)$

$$f_j(\vec{k}_0) := f(\vec{k}_0 + \vec{G}_j). \qquad (8.8)$$

After renaming indices (8.7) becomes

$$D_j(\vec{k}_0, \omega) = \sum_i \varepsilon_{ij}(\vec{k}_0, \omega) E_i(\vec{k}_0, \omega) \qquad (8.9)$$

which establishes a relation between "vectors" E, D by the second-rank tensor ε_{ij}. Do not mix up D, E which are Fourier coefficients of scalars (longitudinal or transverse components of $\vec{D}, \vec{E}$), with vectors $\vec{D}, \vec{E}$ in real space!

Eq. (8.9) tells that Fourier coefficients of D and E whose arguments differ by reciprocal lattice vectors $\vec{G}_j$ are coupled. To put it another way, a long-wavelength external perturbation $D_0(\vec{k}_0, \omega)$ induces polarization charges and hence, via ε, rapidly varying fields E with wave vectors $(\vec{k}_0 + \vec{G}_j)$ in the crystal. Since E oscillates within distances of less than one lattice constant, the induced fields are called local fields [8.1]. They are brought about by the periodic arrangement of charges in the crystal. See Fig. 8.1.

Now the question arises as to how the dielectric tensor ε_{ij} relates to the "macroscopic" dielectric constant defined on the assumption that the medium is homogeneous macroscopically (given by the length scale larger than the magnitude of k^{-1})

$$\bar{D}(\vec{k}, \omega) = \varepsilon_M(\vec{k}, \omega) \bar{E}(\vec{k}, \omega). \qquad (8.10)$$

The bars denote averaging over a lattice period. The averaged displacement field is

$$\bar{D}(\vec{r}, t) = \int\limits_{WS} d^3 r' \, \tilde{D}(\vec{r} - \vec{r}', t) = \frac{1}{(2\pi)^4} \int d\omega \int\limits_{BZ} d^3 k \, D_0(\vec{k}, \omega) e^{i(\vec{k}\vec{r} - \omega t)}.$$

$$(8.11)$$

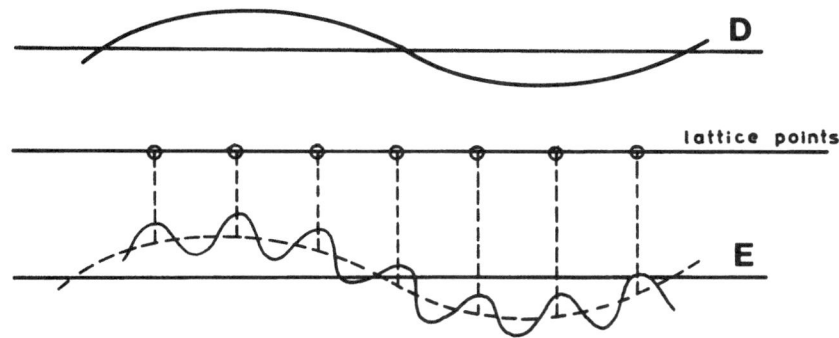

Fig. 8.1. A long-wavelength external perturbation D induces rapidly varying local fields E because charges bound to the periodically arranged atoms of a lattice become polarized.

WS is the Wigner-Seitz cell, and BZ is the first Brillouin zone. Note that only D_0 remains in the integral. All higher harmonics have averaged out.

Hence, the Fourier transforms of $\bar{D}$ and $\bar{E}$ are

$$\bar{D}(\vec{k}_0,\omega) = D_0(\vec{k}_0,\omega),$$
$$\bar{E}(\vec{k}_0,\omega) = E_0(\vec{k}_0,\omega), \tag{8.12}$$

and Eq. (8.10) is

$$D_0(\vec{k}_0,\omega) = \varepsilon_M(\vec{k}_0,\omega)E_0(\vec{k}_0,\omega). \tag{8.13}$$

When a slowly varying external perturbation is applied to the crystal

$$D_i(\vec{k}_0,\omega) \equiv 0 \quad i \neq 0 \tag{8.14}$$

the corresponding electric field can be calculated from the inverse dielectric tensor ε^{-1} as $E = \varepsilon^{-1}D$, or, in components

$$E_i(\vec{k}_0,\omega) = \sum_j \varepsilon_{ij}^{-1}(\vec{k}_0,\omega)D_j(\vec{k}_0,\omega) = \varepsilon_{i0}^{-1}(\vec{k}_0,\omega)D_0(\vec{k}_0,\omega).$$
$$\tag{8.15}$$

Inserting E_0 from Eq. (8.15) in (8.13) we eventually obtain

$$\varepsilon_M(\vec{k}_0,\omega) = 1/\varepsilon_{00}^{-1}(\vec{k}_0,\omega). \tag{8.16}$$

The macroscopic dielectric constant is the inverse of the first diagonal element of the inverse dielectric tensor. This is an important result since intuitively one is tempted to believe that by the averaging process (8.11) all off-diagonal elements of ε_{ij} vanish. Eq. (8.16) shows that this is not the case. Local fields—although they are averaged out in the macroscopic description—show up in the dielectric constant (Otherwise $\varepsilon = \varepsilon_{00}$).

Another consequence of local fields should be mentioned in this context. Plasmons (longitudinal modes) have been defined at the zeros of $\varepsilon(\vec{k},\omega)$ in Chapter 4. When ε is a second-rank tensor, the condition for excitation of plasmons reads $Det(\varepsilon) = 0$, thereby defining *plasmon bands*. In the most simple approximation (the two-band model) the "dielectric matrix" ε_{ij} of virtually infinite order is reduced to a (2x2)-effective matrix, the condition $Det(\varepsilon) = 0$ thus leading to occurrence of two plasmons coupled by Bragg reflection, in a similar way as in the example discussed in Section 6.10. The formerly unperturbed plasmon dispersion graphs split into two lines at the Brillouin zone boundary $G/2$ where they intersect (see Fig. 8.2).

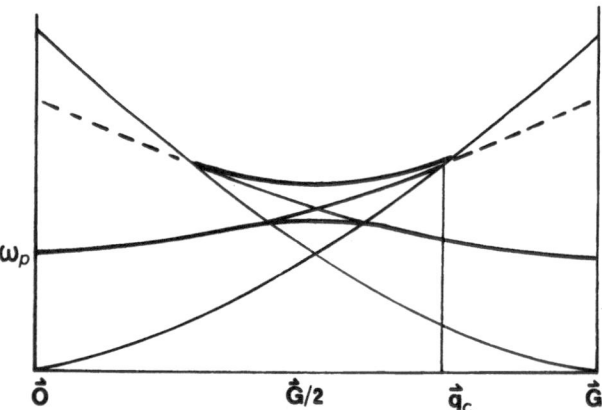

Fig. 8.2. Occurrence of two plasmon branches predicted from interaction with a periodic crystal potential. $\vec{G}$ is a vector of the reciprocal lattice, the plasmon is well defined below q_c. Only the full lines correspond to plasmon maxima.

Since the plasmon is well-defined only for wave numbers smaller than the cut-off wave number q_c at which the plasmon dispersion graph enters the electron-hole continuum (see Fig. 7.1) and hence decays rapidly, two branches should be observable only if

$q_c > G/2$. A candidate crystal where such splitting of a plasmon-like maximum has been observed by inelastic electron scattering is LiF. However, the interpretation given above has not been generally accepted and is still open to discussion [8.2].

8.2. Ion-Electron Interactions

The calculation of ε for a metal, sketched in the previous chapter was based on the jellium model (homogeneous positive background not interacting with electrons). In real metals, ion cores represent localised positive charges. The periodic arrangement of charges in the lattice breaks translational symmetry with the consequences discussed in the previous section. Another consequence is that conduction electrons may now interact with the ions of the lattice, leading to Bragg reflection and to excitation of phonons. Besides that, it can be anticipated that ion-electron interactions will influence the dielectric constant.

As we have seen in Section 7.5, ε is given in the random phase approximation (RPA) as

$$\varepsilon(q,\omega)_{RPA} = 1 + v_q \pi_{RPA}(q,\omega). \tag{7.35}$$

v_q is the Fourier transform of the Coulomb potential. The polarization π, in RPA, is the free electron polarization

$$\pi_{RPA} = \pi_0 \,\hat{=}\, \bigcirc . \tag{8.17}$$

Since the derivation of Eq. (7.35) relied on the jellium model no ion-electron interactions are included in π_{RPA}. This is evidenced by inspection of the graph (8.17).

Ion-electron interaction is described by the potential which the electrons "see" in the crystal

$$V(\vec{r},t) = \sum_i v(\vec{r} - \vec{r}_i(t)), \tag{8.18}$$

where $\vec{r_i}$ are the time-dependent ion positions and v is the potential of a single ion. Eq. (8.18) can be rewritten as

$$V(\vec{r}, t) = \int d^3 r' v(\vec{r} - \vec{r'}) \cdot \sum_i \delta^3(\vec{r'} - \vec{r_i}(t)) \qquad (8.19)$$

The sum over δ-functions is the Fourier transform of the dynamic structure amplitude S [7.2]. The total potential is a convolution integral. In $\vec{q}$-representation, Eq. (8.19) is

$$V(\vec{q}, \omega) = v(\vec{q}) \cdot S(\vec{q}, \omega) \qquad (8.20)$$

and $S(\vec{q}, \omega)$ is the crystal's dynamical structure amplitude.

Graphically, scattering of electrons on the ions via the potential (8.20) can be represented by a vertex as in our example discussed in Section 6.10. Inclusion of all possible sequences of scattering events would then yield the electron's Green function. We shall, however, enter the stage of calculation in a later act and discuss how ion-electron scattering influences the polarization (8.17). In order to proceed we note that any event contributing to the polarization conserves energy and momentum (creation and subsequent destruction of a particle-hole pair). Since a vertex corresponding to the scattering potential (8.20) inserted in the electron propagator means transfer of momentum $\hbar\vec{q}$ and energy $\hbar\omega$ to the ion lattice, there must be at least one compensating scattering event which provides for the formerly lost energy and momentum of the electron. Those processes will be most probable where scattering occurs in pairs [8.3], i.e. the ion lattice acts as a mediator for the exchange of energy and momentum between two electrons, such as does the photon in Coulomb interactions. So, those most probable processes can be considered as two-electron interactions in complete analogy to the Coulomb case. In graphical representation, we denote the fact that the lattice acts as a catalyst for transfer of energy and momentum between electrons by connecting two vertices.

We draw a wavy line:

$\vec{q}$

so as to symbolize interaction of electrons via the potential (8.20).

Each vertex means scattering of an electron. Retaining the RPA for Coulomb $(e^- - e^-)$ interactions, but including ion-electron interactions we have for the polarization

$$\Pi_{RPA} \cong \text{(a)} + \text{(b)} + \text{(c)} + \text{(d)} + \text{(e)} + \text{(f)} + \cdots \cdots \quad (8.21)$$

At present, there is no method of calculation available for the sum total Eq. (8.21). Unlike the series (7.31), divergent terms are not encountered in any order of the perturbation expansion now. We may find a reasonable approximation by truncating the series, say, after term (d), as long as the interaction potential is small. Unfortunately, $v(q)$ is not at all small. Use of a pseudopotential [7.3] $\bar{v}(q)$ which is considerably smaller than $v(q)$ (Fig. 8.3), allows to truncate Eq. (8.21) after term (d).

As a third approximation, we take the static structure amplitude $S(q)$ instead of $S(q,\omega)$ [8.3]. Note that the truncation corresponds to a perturbation expansion in second order of the pseudopotential due to the fact that two scattering processes which factorize as $\bar{v}^2(q)|S(q)|^2$ appear in each of the graphs a) to d). Translation of the truncated series (8.21) by means of Table 6.1 yields then an additional term to the Lindhard expression for the jellium ε. We do not present the formula (which is rather lengthy) but give some results.

Fig. 8.4 is a contour map of the energy loss function of polycrystalline aluminum calculated as discussed above. The quadratic dispersion of the plasmon maximum is clearly visible. Contrary to the Lindhard calculation [7.5] presented in Section 7.5 the plasmon mode has finite energy width for any wave number k. The calculated energy half width (~ 0.4 eV at $k = 0$) is in reasonable agreement with experiment (~ 0.6 eV at $k = 0$ [4.5]).

The dispersion of the plasmon, (the energy of the loss maximum, here drawn as a function of k^2) is a straight line—see Fig. 8.5, where the Lindhard prediction and some experimental results are

152

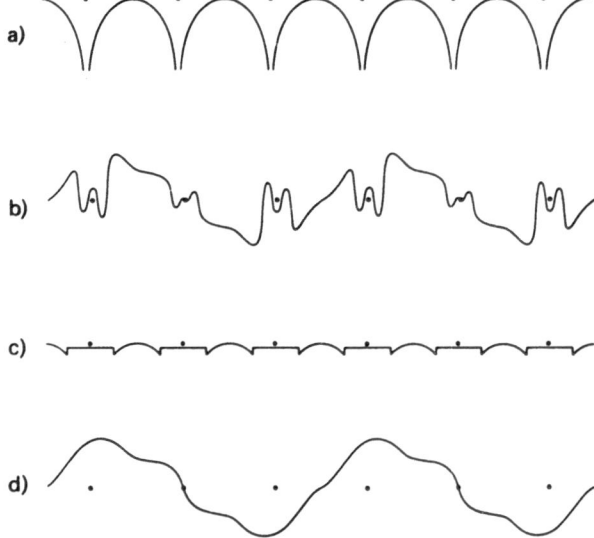

Fig. 8.3. a) ion-core potential, b) corresponding wave function oscillates rapidly at the ion cores, c) pseudopotential, d) corresponding pseudo-wave function is smoother than b). From [7.3].

given for comparison. It is obvious that neither the plasma energy nor the dispersion coefficient is much altered by the inclusion of ion-electron scattering. Other effects, such as interband transitions, seem to be more important in causing deviations from experiment [8.25].

8.3. Multiple Scattering

Repeated scattering of the probe particle traversing a (thin) specimen is encountered in almost all electron probe experiments.

The low-energy parts of Auger or X-ray photoelectron spectra, for instance contain series of maxima due to the multiple excitation of plasmons.

Another—less obvious—example of multiple scattering is comprised by the dynamical diffraction theory which is a description

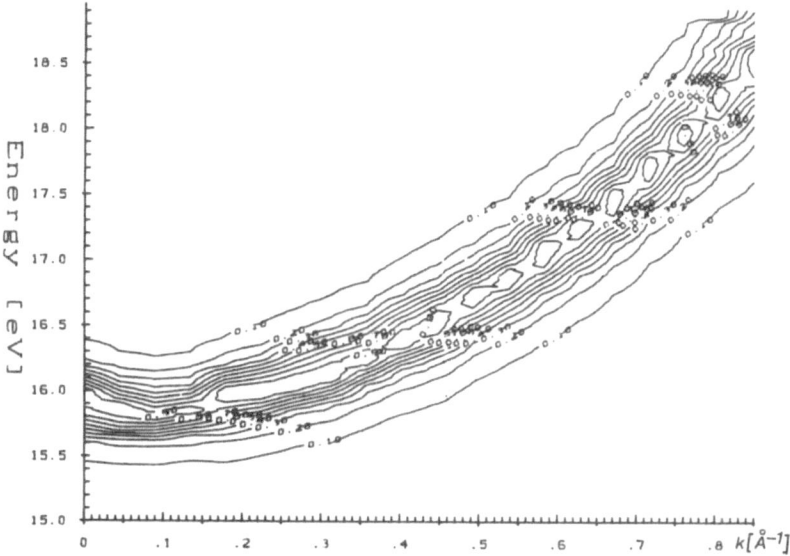

Fig. 8.4. Contour map of the loss function of polycrystalline Al. Perturbation theoretical calculation in second order of the pseudopotential. From [4.5].

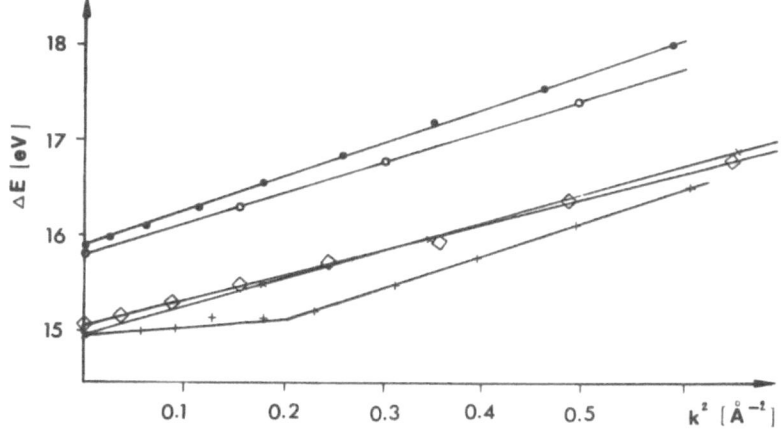

Fig. 8.5. Dispersion diagram of the Al-plasmon, derived from the position of the maximum in Fig. 8.4 (●). Lindhard prediction (○) and experimental results of Kloos (+) [8.4], Batson and Silcox (✗) [8.25] and Urner-Wille (◇) [8.26] are given for comparison.

of how electrons behave in a crystal. Due to repeated scattering of the incident free electron (a plane wave) on lattice planes Bloch

wàves will emerge, with all the well known consequences such as occurence of band gaps, Bragg spots, channeling or thickness fringes
[8.5]. We have discussed this effect in Chapter 6 in a much simplified form: Expression (6.57) for the perturbed propagator shows
the subsequent scattering processes graphically.

It is well known that in single crystals multiple subsequent inelastic and elastic scattering is responsible for the formation of
Kikuchi lines in the diffraction pattern. A good description of multiple scattering is also important for a more accurate calculation of
what is known as "absorption coefficients" in electron microscopy.
By absorption the fact is denoted that inelastically scattered electrons may not pass through the objective aperture and therefore
cannot contribute to the image formation.

An important issue recently discussed by Rez [8.6] is the spatial
resolution in anlytical electron microscopy which is limited by the
broadening of the probe traversing the specimen. Beam spreading
is induced by small angle scattering [8.7] with multiple processes
strongly contributing even in the case of moderate thickness (Fig.
8.6).

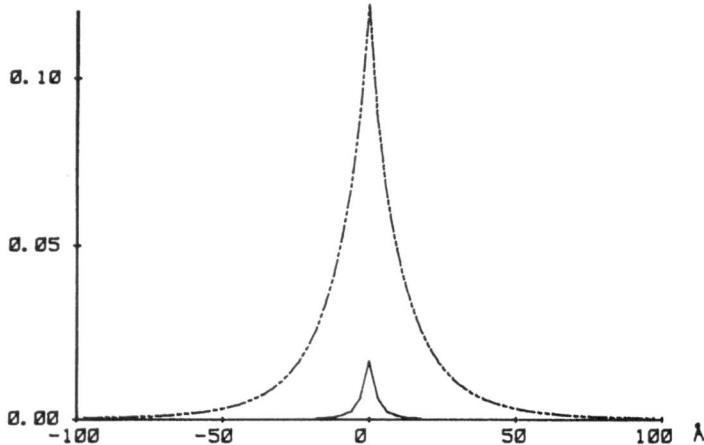

Fig. 8.6. Normalized spatial distribution of electron current density emerging
from an iron specimen 50 nm thick (D=4.6) assuming single scattering (full
line) compared to full inclusion of multiple scattering. From [8.6].

In treating multiple scattering one has to distinguish between
coherent and incoherent subsequent scattering processes. Coher-

ent elastic interactions are well described within the framework of dynamical diffraction theory (see, e.g. [8.5]). Yoshioka [8.8], and Serneels and Haentjens [8.9] have extended the dynamical theory to inelastic scattering.

It is probably because of the deterrent formalism that this approach has been restricted in the last decades to rather theoretical treatments which are only marginally useful in practice [8.10], [8.11], [8.12].

Some interesting approaches deal with the combined effects of elastic scattering and electronic excitations [8.27], plasmons [8.28], or phonons [8.29]. Though, a unified description of multiple elastic-inelastic scattering with emphasis on practical aspects such as formation of Kikuchi bands is still missing. It is certainly beyond the scope of this monograph to discuss any of these issues. The interested reader is referred to the book of Ohtsuki [8.13] for a recent review of multiple coherent scattering.

The incoherent case has been approached quite differently. A number of methods have been proposed so as to deal with inelastic multiple scattering. The various approaches can be grouped into those which aim at calculating from given scattering cross sections the distribution of electrons having passed the specimen; and into solutions of the inversion problem, which means quite the contrary, viz. determination of scattering cross sections from the measured distribution of electrons having passed the specimen.

Historically, the former methods were developed much earlier. W. Bothe, as early as 1929, presented a solution of the problem based on the transport equation for electrons in matter [8.14]. Later on a number of authors reexamined the problem [8.15], [8.16]. Recent examples demonstrate that the item has not yet been fully resolved [8.6]. The proliferation of computers rendered possible the numerical simulation and tracing of a large number of arbitrarily selected electron trajectories in matter, an approach to the prediction of spatial and directional distribution of electrons known as Monte Carlo (MC) simulation and widely used today. As an example MC calculations of electron trajectories are presented in Fig. 8.7.

The major drawback of this category is its indirect access to physically relevant data: In order to learn of. say, the single scat-

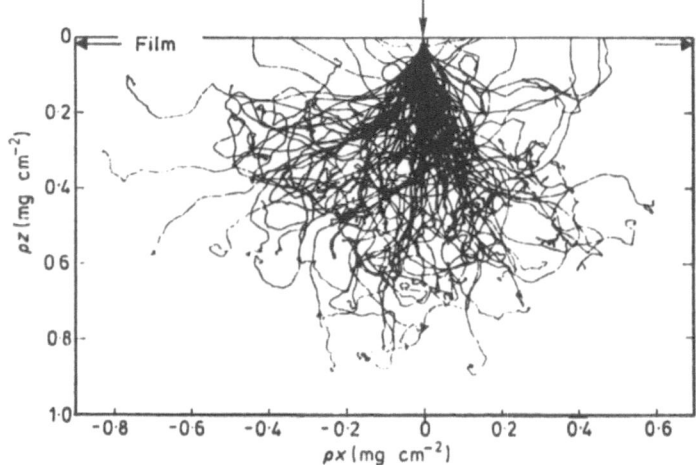

Fig. 8.7. MC simulation of trajectories of electrons entering a specimen from top. From [8.17].

tering cross section, one has to assume a particular cross section, then calculate the directional distribution of electrons leaving the specimen on this assumption, and compare the result with experiments.

The second category of previously mentioned methods aim at extracting scattering cross sections from measured profiles directly. Any such approach is a solution of the *inversion problem* so called because the physical phenomenon of multiple scattering leading to profile broadening is mathematically inverted. Most inversion methods were originally designed for the case of inelastic, incoherent scattering and did not account for the influence of the scattering angle. An appropriate experimental situation for this case is depicted in Fig. 8.8.

In the following, we shall derive, for the experimental setup of Fig. 8.8, a relation between the single scattering cross section and the multiple scattering spectrum. A specimen will be irradiated homogeneously over a distance considerably larger than its thickness. On the assumption that only small angle scattering occurs (which is approximately valid for electrons of some keV energy or more), the probability Q_n that an electron has been scattered n-fold no longer depends on the scattering angle or the in-plane coordinates

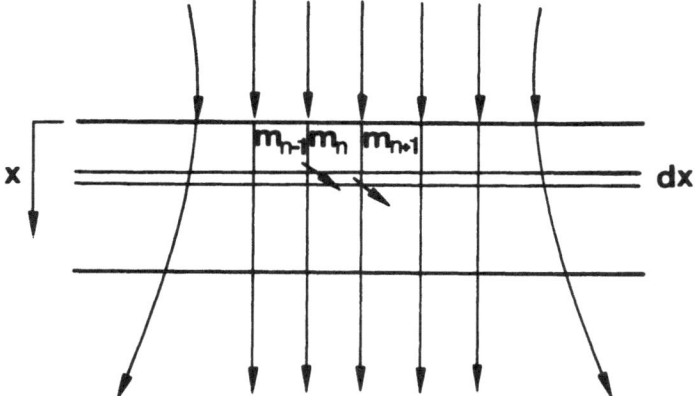

Fig. 8.8. Thin specimen, irradiated homogeneously.

but on the depth x in the specimen only.

At depth x we find a number $m_n(x)$ of electrons scattered n-fold. The change of m within dx is given, by a local balance argument, —see full arrows in Fig. 8.8— as

$$dm_n = (m_{n-1} - m_n)\alpha dx, \qquad (8.22)$$

in which α is a constant describing the rate of scattering. From Eq. (8.22), we have a set of differential equations for m_n, or, scaling by the sum total of electrons, for probabilities Q_n:

$$\dot{Q}_n(x) = \alpha(Q_{n-1} - Q_n) \qquad (8.23)$$

the solution of which is

$$Q_n = \frac{(x\alpha)^n}{n!} e^{-\alpha x} \qquad (8.24)$$

as can be verified by inserting (8.24) into (8.23). Putting $n = 0$, the well known exponential attenuation of the "primary beam" is regained. Comparison with Eq. (1.8) in the following form

$$I_0(x) = I_0(0)e^{-\sigma n x} \qquad (8.25)$$

shows that α equals the inverse mean free path:

$$\alpha = \sigma n = \lambda^{-1}. \qquad (8.26)$$

When we agree to measure the depth x in units of λ, now denoted D

$$D = x/\lambda, \tag{8.27}$$

we obtain, from Eq. (8.24), for the n-fold scattering probability Q_n as a function of the dimensionless thickness of the specimen, Poisson's distribution:

$$Q_n = e^{-D} D^n / n! \tag{8.28}$$

Q_0 is the probability that the particle was *not* scattered.

It is evident that

$$P = \sum Q_n = 1. \tag{8.29}$$

In case of angle-resolved inelastic scattering it is convenient to define normalized scattering functions $f_1(E, \Omega)$ of solid angle and energy loss E such that

$$\int_{4\pi} d\Omega \int dE f_1(E, \Omega) = 1. \tag{8.30}$$

In spherical coordinates, Ω is short for (ϑ, φ), ϑ is the scattering angle with respect to the incident beam, φ is the azimuth.

The normalized n-fold scattering function f_n is then given as a convolution integral

$$f_n(E, \vartheta, \varphi) = \int_0^E \int_{4\pi} f_{n-1}(E', \vartheta', \varphi') f_1(E - E', \vartheta'', \varphi'') d\Omega' dE' \tag{8.31}$$

in which the vector $\Omega'' = (\vartheta'', \varphi'')$ is given by $\Omega'' = \Omega - \Omega'$. The case $n = 1$ in equation (8.31) defines

$$f_0(E, \vartheta, \varphi) = \delta(E, \vec{\Omega}), \tag{8.32}$$

the normalised zero-scattering function, i.e. a monoenergetic, parallel incident beam. From Eqs. (8.30) and (8.31) it follows that f_n are normalized in the same sense as was f_1.

The differential probability p for scattering with a particular energy loss E and a particular scattering angle (ϑ, φ) is a weighted sum over all f_n

$$p(E, \vartheta, \varphi) = \sum_{n=0}^{\infty} q_n f_n(E, \vartheta, \varphi). \qquad (8.33)$$

From (8.30) and (8.31)

$$Q_n = \int q_n f_n(E, \vartheta, \varphi) dE d\Omega = q_n, \qquad (8.34)$$

and (8.28) is

$$q_n = e^{-D} D^n / n!. \qquad (8.35)$$

Equation (8.31), repeatedly inserted in (8.33), together with (8.35) constitute an integral equation which has to be solved for the normalized single scattering function $f_1(E, \vartheta)$, from a measured differential probability p. For radial symmetry (as is the case in isotropic media or randomly oriented crystallites), all probabilities are functions of only two independent variables E, ϑ.

The Fourier transform of a convolution, such as Eq. (8.31) yields a product which is easier to manipulate. In order to simplify Eq. (8.31) in this sense one has to use a system of orthogonal functions defined in $[0, \pi]$ with respect to which the scattering functions $f_n(\vartheta)$ can be Fourier transformed. The Legendre polynomials $P_l(\cos \vartheta)$ meet these requirements. Any piecewise continuous function can be expanded as

$$f_n(\vartheta) = \sum_{l=0}^{\infty} \frac{2l+1}{4\pi} a_l^{(n)} P_l(\cos \vartheta), \qquad (8.36)$$

$$a_l^{(n)} = \int d\Omega f_n(\vartheta) P_l(\cos \vartheta). \qquad (8.37)$$

It should be noted that Legendre-coefficients usually contain a factor $(2l+1)/4\pi$. For later convenience, we have put this factor into the series Eq. (8.35).

By means of the orthogonality of P_l and the shift theorem of spherical harmonics it can be shown that (8.31), (8.33) and (8.35) yield

$$a_l^{(n)}(E) = \int dE' a_l^{n-1}(E') a_l^{(1)}(E - E'), \qquad (8.38)$$

$$c_l(E) = e^{-D} \sum_{n=0}^{\infty} \frac{D^n}{n!} a_l^{(n)}(E) \qquad (8.39)$$

where

$$c_l(E) = \int d\Omega p(E, \vartheta) P_l(\cos \vartheta). \qquad (8.40)$$

Whereas in the original Eq. (8.31) and (8.33) the single scattering function f_1 at *all* energies and angles contributed to p at a particular E, ϑ, in Eq. (8.39) it is only *one* expansion coefficient a_l at *all* energies which contributes to c. Thus the coupling between different angles has been eliminated by means of the generalized Fourier transform. We have obtained a set of decoupled integral equations for the expansion coefficients of $f_1(E, \vartheta)$.

Equations (8.38) and (8.39) no longer contain the scattering angle. They are formally equivalent with the equations for multiple scattering in the angle-integrated formulation [8.18], [8.19], and may be solved by any of the methods which work in that case. We shall rely on a method developed by the author [8.20]. In the paper cited above it has been shown that the integral equation of type (8.38) and (8.39) can be replaced by

$$\mathbf{C}_l = \exp[\mathbf{D}(\mathbf{A}_l - \mathbf{1})] \qquad (8.41)$$

where the entries of the matrices $\mathbf{A}, \mathbf{C}$ are defined respectively as

$$A_{l,ij} = \int_{E_{i-j}-\Delta E/2}^{E_{i-j}+\Delta E/2} dE a_l^{(1)}(E) \qquad (8.42)$$

and

$$C_{l,ij} = \int_{E_{i-j}-\Delta E/2}^{E_{i-j}+\Delta E/2} dE c_l(E). \qquad (8.43)$$

Here, $E_0 := 0$, and ΔE is the equidistant spacing between subsequent values E_i. Eq. (8.41) is approximately equivalent to (8.38) and (8.39), the approximation error tending to zero when $\Delta E \to 0$. We note that in this limit equation (8.41) is a set of nonlinear operator equations with continuous convolution operators A_l^* defined as $A_l^* f := a_l * f$.

Apart from a factor $\propto \Delta E$, the entries of $\mathbf{C}$ are the generalized Fourier coefficients of an angle-resolved energy loss profile measured at energy E_{i-j}.

In principle, matrices $\mathbf{C}, \mathbf{A}$ are of infinite order. However, (8.41) holds for any finite diagonal submatrix containing $C_{l,00}$, as can be shown by partitioning [8.20]. The task of solving (8.41) is accomplished by

$$\mathbf{A}_l = D^{-1} \ln \mathbf{C}_l + \mathbf{1} \tag{8.44}$$

the logarithm to be understood in terms of a power series in $\mathbf{C}_l$

$$\mathbf{A}_l = D^{-1}[(\mathbf{C}_l - \mathbf{1}) - \frac{1}{2}(\mathbf{C}_l - \mathbf{1})^2 + \frac{1}{3}(\mathbf{C}_l - \mathbf{1})^3 - \ldots]. \tag{8.45}$$

$\mathbf{1}$ is the unity matrix of same order as $\mathbf{C}_l$.

Elastic small angle scattering is covered by the present formalism as the limiting case

$$f_1(E, \vartheta) = g_1(\vartheta)\delta(E). \tag{8.46}$$

Inserting (8.46) in (8.37) for $n = 1$ and observing (8.42) yields

$$\mathbf{A}_l = \int d\Omega g_1(\vartheta) P_l(\cos\vartheta) \cdot \mathbf{1} = a_l \cdot \mathbf{1}, \tag{8.47}$$

i.e. the matrix Eq. (8.44) degenerates to a scalar equation,

$$a_l = D^{-1} \ln c_l + 1. \tag{8.48}$$

Results of the retrieval of single scattering profiles are shown in Fig. 8.9 for samples of thickness $D = 4$ and $D = 8$. The corresponding multiple scattering profiles are shown in the figure, too. The subsidiary peak in Fig. 8.9c is caused by numerical errors in the retrieval procedure. Note, however, that the case $D = 8$ is a rather exotic one hopefully never encountered in practice.

Another important case is the retrieval of single inelastic scattering profiles from transmission energy loss spectra. Such spectra are mainly obtained in the electron microscope working in the image mode [8.21]. We assume that the aperture of the imaging system is large enough so as to collect almost all of the electrons scattered inelastically through different angles at least for a reasonable specimen thickness, i.e. some mean free path lengths. Consequently, the measured profile may be regarded as the integral of $p(E, \vartheta)$ over the solid angle 4π.

Referring to Eq. (8.40), the profile is given by $C_0(E)$, and, due to Eq. (8.36) the corresponding single scattering function $f_1(E)$ is proportional to $a_0(E)$. Hence, to calculate single scattering loss probabilities, one simply resorts to $l = 0$ in Eq. (8.45).

This formula was derived previously by the author for the particular case of angle-integrated measurements [8.20] and has been shown to be superior to the Fourier transform, which is usually applied [8.22]. According to the fact that energy loss spectra are aperiodic, any finite base interval chosen for the Fourier expansion must lead to a misrepresentation. Satisfactory results are obtained only when data are sampled up to an energy where the intensity practically vanishes [8.18]. See Fig. 8.10 for retrieval by means of Fourier transforms.

In application, spurious oscillations often appear in the processed spectra as is visible at the origin in Fig. 8.10b and at the multiple plasmon losses in Fig. 8.11. They are due to the enhancement of statistical noise by deconvolution, to the drift and inaccuracy of the zero loss, to truncation of the loss range, and to inaccuracies in the specimen thickness.

Fig. 8.12 shows results of the matrix approach, in comparison to the Fourier method. Note that in this case the spectrum was truncated immediately after the "single scattering" peak without influence on the retrieval result, whereas the Fourier method induced spurious oscillations.

Solutions of the general case (angle-resolved multiple inelastic scattering) have been calculated assuming a sharply peaked single scattering function and simulating the angle-resolved energy loss

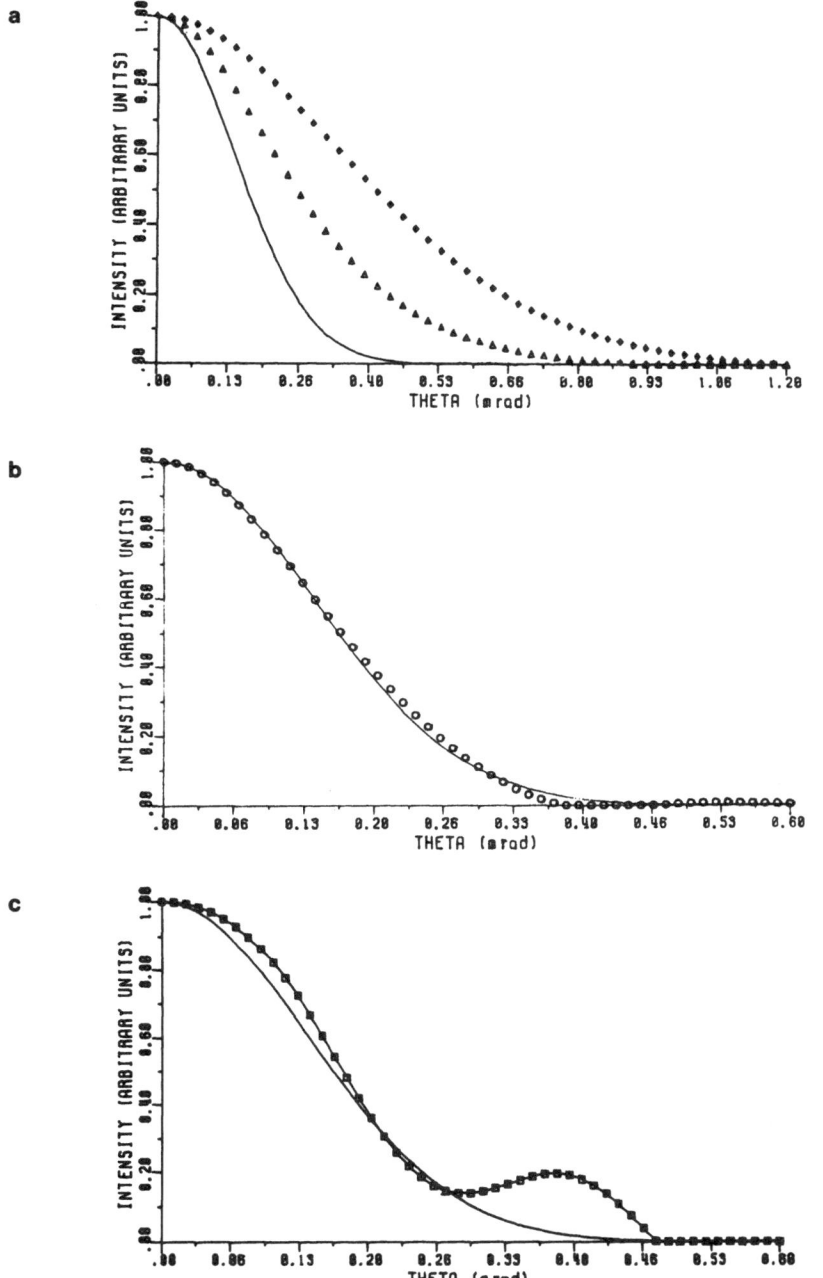

Fig. 8.9. (a) Simulated multiple scattering profiles, for D=4 (Δ) and D=8 ($\diamond$), (b) single scattering function $g_1(\vartheta)$ retrieved from (a) for D=4. (c) same as in (b) for D=8. The full curve is the single scattering input $g_1(\vartheta)$. From [8.23].

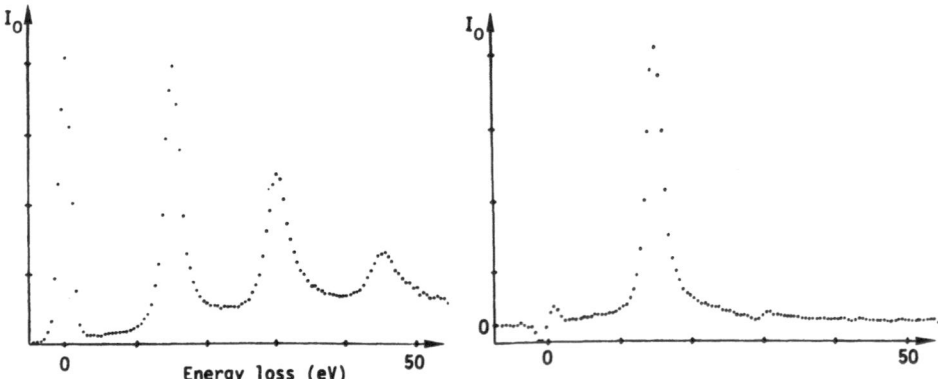

Fig. 8.10. (a) Experimental energy loss spectrum recorded from an aluminum foil at 100 kV accelerating voltage. A portion only of the data is shown which was collected out to 180 eV loss, where the spectrum amplitude falls to less than 1 % of the maximum. (b) The single loss spectrum retrieved from Fig. (a) using a Fourier transform . The spurious peak at the energy origin results from a small peak shift in the instrumental impulse response and affects the spectrum only at the origin. The peak at 15 eV is the plasmon loss. Subsidiary maxima are caused by multiple scattering. From [8.18].

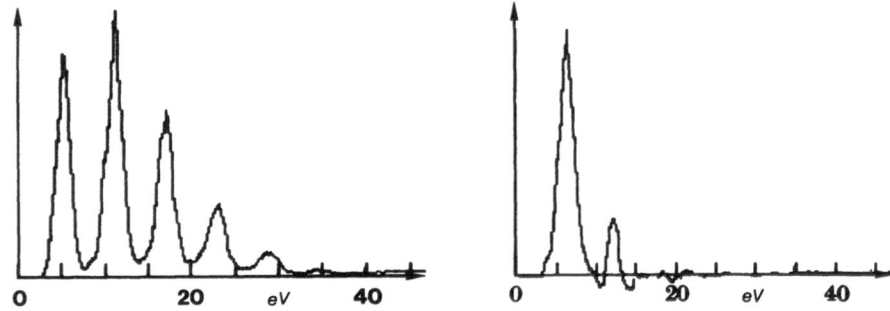

Fig. 8.11. (a) Plasmon-loss spectrum of sodium.
(b) Spectrum deconvoluted by using an approximate Fourier-transform method; the zero-loss peak has not been added back in. From [8.24].

profile (Fig. 8.13a). The retrieval procedure practically eliminates the multiple losses completely (Fig. 8.13b). Within the tolerance of drawing, the results coincide with the simulation input.

Altogether, it is concluded that the Fourier transform is suited for routine investigation, due to the simple and well established formalism. However, if more accurate spectra are demanded (e.g. for Kramers-Kronig-analysis), or in case of angle-resolved experiments, the matrix approach seems to be more appropriate. Whether the

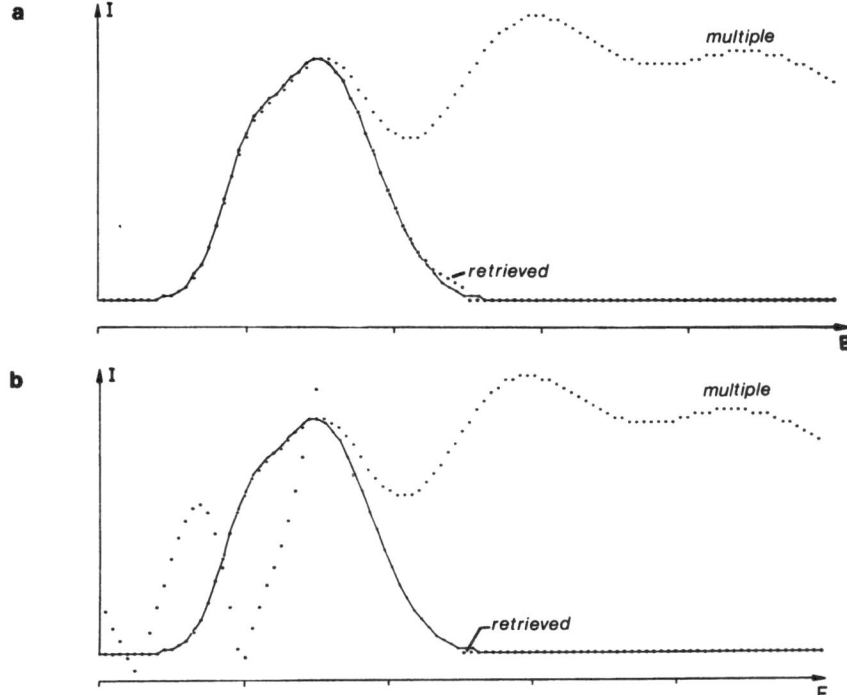

Fig. 8.12. Retrieval of energy loss functions. (a) Multiple scattering matrix algorithm, sixty terms of the series expansion Eq. (8.45). Multiple scattering spectrum (+) simulated from a hypothetical single scattering function (full line). Retrieval (*) by use of matrix algorithm.
(b) Same as in Fig. (a), retrieval (*) by use of the Fourier algorithm. Values rise beyond scale at the maximum.

latter is sensitive to noise of experimental inaccuracies remains to be seen and will be determined after more extensive numerical analysis has been performed.

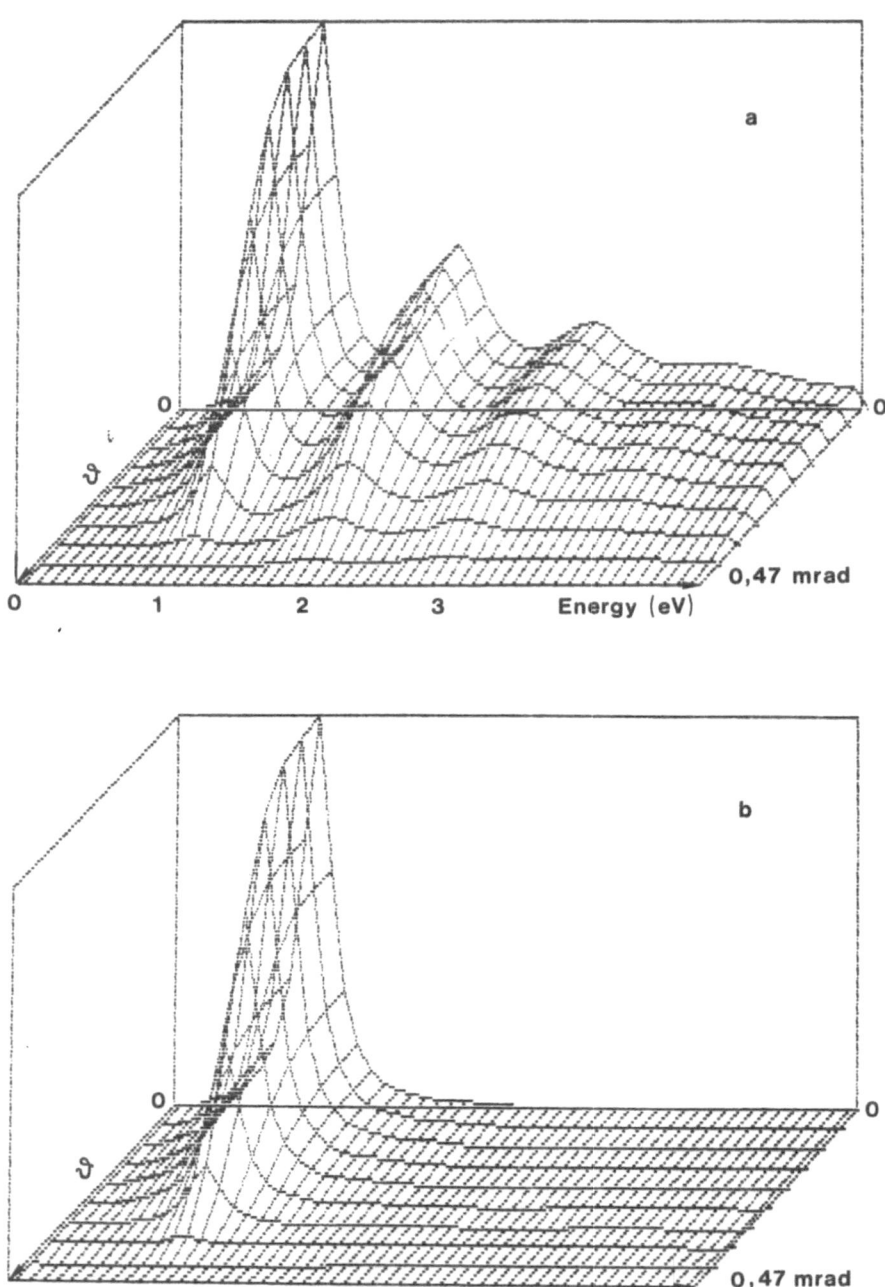

Fig. 8.13. (a) Angle-resolved multiple loss profile and (b) single loss retrieved by Eq. (8.45).

References

Chapter 1 Classical Scattering Theory

1.1   Rutherford E (1911) Phil Mag 21, 669
1.2   Sexl R, Sexl H (1981) Weiße Zwerge — Schwarze Löcher. Vieweg, Braunschweig
1.3   Johnson RE (1982) Introduction to Atomic and Molecular Collisions. Plenum, New York
1.4   Massey H (1979) Atomic and Molecular Collisions. Taylor & Francis, London
1.5   Mott NF, Massey H (1965) The Theory of Atomic Collisions. Clarendon Press, Oxford
1.6   Williams BG, Egerton RF (1982) Chem Phys Lett 88, 95
1.7   Ramsauer C (1921) Ann d Phys 64, 513
1.8   Townsend PD, Bailey P (1922) Phil Mag 43, 593
1.9   Townsend PD, Bailey P (1922) Phil Mag 44, 1033

Chapter 2 Quantum Mechanical Scattering Theory

2.1   Bethe H (1930) Ann Phys 5, 325
2.2   Sommerfeld A, Bethe H (1967) Elektronentheorie der Metalle. Springer, Berlin
2.3   Fick E (1968) Einführung in die Grundlagen der Quantentheorie. Akad. Verlagsges, Leipzig
2.4   Platzmann PM, Wolf PA (1973) Waves and Interactions in Solid State Plasmas. Solid State Phys Suppl 13. Academic Press, New York
2.5   Van Hove L (1954) Phys Rev 95, 249
2.6   Baltes HP (ed.) (1980) Topics in Current Phys 20. Springer, Berlin
2.7   Inokuti M (1971) Rev Mod Phys 43, 297
2.8   Moiseiwitsch BL (1968) Rev Mod Phys 40, 238

2.9   Fano U, Cooper JW (1968) Rev Mod Phys 40, 441

2.10   Egerton RF (1979) Ultramicroscopy 4, 169

2.11   Landau LD, Lifschitz EM (1979) Lehrbuch der Theoretischen Physik III. Akademie-Verlag, Berlin Ost

2.12   Herman F, Skillman S (1963) Atomic structure calculations. Prentice Hall, New York

2.13   Manson ST (1972) Phys Rev A 6, 1013

2.14   Ritsko JJ, Schnatterly SE, Gibbons PC (1974) Phys Rev Lett 32, 671

2.15   Ferch J, Granitza B, Raith W (1985) J Phys B 18, L445

2.16   McCarthy IE, Mitroy JD, Stelbovics AT (1985) J Phys B18, 2509

2.17   Abramowitz M, Stegun IA (eds.) (1972) Handbook of mathematical functions. Dover Publ, New York

2.18   Einstein A (1910) Ann Physik (Leipzig) 33, 1275

2.19   Egerton RF (1981) Ultramicroscopy 6, 297

2.20   Vradis A, Priftis GD (1985) Phys Rev B32, 3556

Chapter 3 Practical Aspects of Absorption Edge Spectrometry

3.1a   Egerton RF (1981) Ultramicroscopy 7, 207

3.1b   Joy DC, Maher DM (1981) J Microsc 124, 37

3.2   Colliex C (1982) J Microsc Spectrosc Electron 7, 525

3.3   Williams BG (ed.) (1977) Compton Scattering. McGraw-Hill, New York

3.4   Williams BG, Sparrow TG, Egerton RF (1984) Proc Roy Soc London A 393, 409

3.5a   Williams BG, Thomas JM (1983) Internat Rev Phys Chem 3, 39

3.5b   Weyrich W, Pattison P, Williams BG (1979) Chem Phys 41, 271

3.6a   Tafto J, Krivanek OL (1982) Phys Rev Lett 48, 560

3.6b   Tafto J, Krivanek OL (1982) Nuclear Instr Meth 194, 153

3.7   Leapman RD, Cosslett VE (1976) J Phys D 9, L29

3.8a   Egerton RF (1981) Proc of the 39. EMSA-Meeting, 198

3.8b Egerton RF (1981) J Microsc 123, 333

3.9   Joy DC, Maher DM (1980) Ultramicroscopy 5, 333

3.10 Isaacson M, Johnson D (1975) Ultramicroscopy 1, 33

3.11 Eisenberger P, Platzmann PM (1970) Phys Rev A 2, 415

3.12 Bauer GEW, Schneider JR (1983) Solid State Communications 47 (9), 673

3.13 Hirsch P, Howie A, Nicholson RB, Pashley DW, Whelan MJ (1977) Electron Microscopy of thin Crystals. Krieger, New York

3.14 Williams BG, Bourdillon AJ (1982) J Phys C: Solid State Phys 15, 6881

3.15 Leapman R (1984) Electron Beam Interactions with Solids. AMF O'Hare, Chicago

3.16 Brown FC (1974) Sol State Phys 29, 1

3.17 Economou EN (1979) Green's Functions in Quantum Physics. Springer Series in Solid State Phys 7, Springer, Berlin

3.18 Colliex C, Trebbia P (1982) Ultramicroscopy 9, 259

3.19 Colliex C, Krivanek OL, Trebbia P (1981) Inst Phys Conf Ser 61, 183

3.20 Cazaux J (1983) Ultramicroscopy 12, 83

3.21 Hitchcock AP, Teng CH (1985) Surf Sci 149, 558

3.22 Egerton RF (1982) Phil Trans R Soc London A 305, 521

3.23 Wong J (1981) Topics in Applied Physics 46, 45

3.24 Egerton RF (1980) Instrumentation and Software for Electron Energy Loss Microanalysis. In: Scanning Electron Microscopy. AMF O'Hare, Chicago

3.25 Silcox J (1979) Ultramicrosc 3, 409

3.26 Johnson DE (1984) Electron Energy Loss Spectrometry. In: Echlin P (ed.) Analysis of Organic and Biological Surfaces. Wiley & Sons, New York

3.27 Disko MM, Spence JCH, Sankey OF, Saldin D (1986) Phys Rev B33, 5642

3.28 Egerton RF (1986) Electron Energy Loss Spectroscopy in the Electron Microscope. Plenum Press, New York, London

170

Chapter 4 Electrodynamics in Homogeneous, Isotropic Media

4.1 Langmuir I (1926) Proc Nat Acad Sci 14, 627
4.2 Tonks L, Langmuir I (1929) Phys. Rev 33, 195, 996
4.3 Venghaus H (1975) phys stat sol (b) 71, 609
4.4 Ehrenreich H, Philipp HR (1962) Phys Rev 128, 1622
4.5 Sölkner G (1986) Plasmonen in einfachen Metallen. Thesis, Technical University Vienna
4.6 Daniels J (1971) Optics Comm 3, 240
4.7 Wehenkel C, Gauthé B (1974) phys stat sol (b) 64, 515
4.8 Jackson JD (1975) Classical Electrodynamics. John Wiley & Sons, New York, London, Sydney, Toronto
4.9 Jonscher AK (1980) Phys Thin Films 2, 205
4.10 Raether H (1980) Excitation of Plasmons and Interband Transitions by Electrons. Springer Tracts in Modern Physics 88. Springer, Berlin, Heidelberg, New York
4.11 Kittel C (1966) Introduction to Solid State Physics. John Wiley & Sons, New York
4.12 Sommerfeld A (1977) Vorlesungen über Theoretische Physik (Elektrodynamik). Harri Deutsch, Thun, Frankfurt/M.
4.13 Agranovich VM, Galanin MD (1982) Electronic Excitation Energy Transfer in Condensed Matter. In: Agranovich VM, Maradudin AA (eds) Modern Problems in Condensed Matter Sciences. North-Holland, Amsterdam, New York, Oxford
4.14a Ruthemann G (1941) Naturwissenschafften 29, 648
4.14b Ruthemann G (1948) Ann Phys 2, 113
4.15 Lang W (1948) Optik (Stuttgart) 3, 233
4.16 Boltzmann L (1893) Vorlesungen über Maxwells Theorie der Elektrizität und des Lichts. München
4.17 Behmer M, Claus R (1984) Phys Rev B30, 4800

Chapter 5 Some Details on Charge Oscillations

5.1 Debye PP, Hückel E (1923) Phys Z 24, 185
5.2 Friedel J (1958) Nuovo Cimento Suppl 7, 287
5.3 Kohn W (1959) Phys Rev Lett 2, 393
5.4 Kohn W, Vosko SH (1960) Phys Rev 119, 912

5.5 Robusto PF, Braunstein R (1981) phys stat sol (b) 107, 443

5.6 Boardman AD (1982), Hydrodynamic Theory of Plamon-Polaritons on Plane Surfaces. In: Boardman AD (ed.) Electromagnetic Surface Modes. John Wiley & Sons, Sussex

5.7 Glicksman M (1971) Plasmas in Solids. In: Ehrenreich H, Seitz F, Turnbull D (eds.) Solid State Physics, Vol. 26, Acad. Press, New York

5.8 Lopez-Rios T, Vuye G (1982) J Phys E: Sci Instrum 15, 456

5.9 Mathewson AG, Myers HP (1971) Phys Scr 4, 291

5.10 Otto A (1968) Z Phys 216, 398

5.11 Lehner G (1975) Fusionsexperimente. In: Gobrecht H (ed.) Bergmann-Schäfer—Lehrbuch der Experimentalphysik, Bd. IV, 2. W de Gruyter, Berlin, New York

5.12 Forstmann F, Gerhardts RR (1982) Festkörperprobleme XXII, 291

5.13 Sauter F (1967) Z Phys 203, 488

5.14 Boardman AD, Ruppin R (1981) Surface Sci 112, 153

5.15 Langkowski J (1975) J Phys D: Appl Phys 8, 2058

5.16 Raether H (in press) Surface Plasmons on Smooth and Rough Interfaces and on Gratings. Springer, Berlin, Heidelberg, New York.

5.17 Forstmann F, Gerhardts RR (in press) Metal Optics near the Plasma Frequency. Springer Tracts in Modern Physics 109. Springer, Berlin, Heidelberg, New York

Chapter 6 Quantum Mechanical Preliminaries

6.1 Demidovich BP, Maron IA (1973) Computational Mathematics. Mir Publishes, Moscow

6.2 Courant R, Hilbert D (1968) Methoden der mathematischen Physik. Springer-Verlag, Berlin, Heidelberg, New York

6.3 Raimes S (1972), Many-Electron Theory. North-Holland, Amsterdam, London

6.4 Mattuck RD (1967) A Guide to Feynman Diagrams in the Many-Body Problem. McGraw-Hill, London

Chapter 7 Quantum Mechanical Description of the Electron Gas

7.1 Fetter AL, Walecka JD (1971) Quantum Theory of Many-Particle Systems. McGraw-Hill, New York
7.2 Pines D (1964) Elementary Excitations in Solids. W.A. Benjamin, New York, Amsterdam
7.3 Madelung O (1972) Festkörpertheorie. Springer-Verlag, Berlin, Heidelberg, New York
7.4 Pines D (1955) in: Seitz F, Turnbull D (eds.) Advances in Solid State Physics. Academic Press, New York
7.5 Lindhard J (1954) Kgl. Danske Videnskab. Selskab, Mat-fys. Medd 28, 8

Chapter 8 Beyond Simple Models

8.1 Adler SL (1962) Phys Rev 126, 413
8.2 Schnatterly S (1983) Determination of $S(q,\omega)$ by Inelastic Electron and X-ray Scattering. In: Devreese JT, Brosens F (eds.) Electron Correlations in Solids, Molecules, and Atoms. Plenum Press, New York, London
8.3 Helman JS, Baltensperger W (1966) Phys Kondens Materie 5, 60
8.4 Kloos T (1973) Z Phys 265, 225
8.5 Spence JCH (1981) Experimental High-Resolution Electron Microscopy. Clarendon Press, Oxford
8.6 Rez P (1983) Ultramicroscopy 12, 29
8.7 Misell DL, Burge RE (1969) J Phys C2, 61
8.8 Yoshioka H (1957) J Phys Soc Japan 12, 6, 618
8.9 Serneels R, Heantjens D (1980) Phil Mag A42, 1
8.10 Yamamoto T (1980) Acta Crystallogr A36, 126
8.11 Gjønnes J (1966) Acta Crystallogr 20, 240
8.12 Chukovskii FN, Alexanjan LA, Pinsker ZG (1973) Acta Crystallogr A29, 38
8.13 Ohtsuki Y-H (1983) Charged Beam Interactions with Solids. Taylor & Francis Ltd. London, New York

8.14   Bothe W (1929) ZS Physik 54, 161

8.15   Landau LD (1944) J Phys USSR 8, 201

8.16   Goudsmit S, Saunderson JL (1940) Phys Rev 57, 24

8.17   Armigliato A, Desalvo A, Rinaldi R, Rosa R (1979) J Phys D: Appl Phys 12, 1299

8.18   Spence JCH (1979) Ultramicroscopy 4, 9

8.19   Misell DL, Jones AF (1969) J Phys A2, 540

8.20   Schattschneider P (1983) Phil Mag B47, 555

8.21   Johnson DE (1979) Ultramicroscopy 3, 361

8.22   Schattschneider P, Sölkner G (1984) J Microscopy 134, 73

8.23   Schattschneider P, Zapfl M, Skalicky P (1985) Inverse Problems 1, 381

8.24   Egerton RF, Williams BG, Sparrow TG (1985) Proc R Soc London A398, 395

8.25   Batson PE, Silcox J (1983) Phys Rev B27, 5224

8.26   Urner-Wille M, Raether H (1976) Phys Lett 58A, 265

8.27   Kainuma Y (1955) Acta Crys 8, 247

8.28   Howie A (1963) Proc Roy Soc London 271 A, 268

8.29   Takagi S (1958) J Phys Soc Japan 13, 278

Author Index

Abramowitz M 36
Adler SL 146
Agranovich VM 82
Alexanjan LA 155
Armigliato A 156
Bailey P 12
Baltensperger W 150
Baltes HP 21
Batson PE 153
Bauer GEW 48
Behmer M 88
Bethe H 15
Boardman AD 85, 92
Boltzmann L 62
Bothe W 155
Bourdillon AJ 50
Braunstein R 87
Brown FC 43
Burge RE 154
Cazaux J 41
Chukovskii FN 155
Claus R 88
Colliex C 41
Cooper JW 26
Cosslett VE 118
Courant R 104
Daniels J 56
Debye PP 76
Demidovich BP 103
Desalvo A 156
Disko MM 54
Egerton RF 11, 15, 26,
    42, 44, 54, 164
Ehrenreich H 55

Einstein A 20
Eisenberger P 48
Fano U 26
Ferch J 40
Fetter AL 123
Fick E 16
Forstmann F 89
Friedel J 77
Galanin MD 82
Gauthé B 56
Gerhardts RR 89
Gibbons PC 33
Gjønnes J 155
Glicksman M 79
Goudsmit S 155
Granitza B 40
Haentjens D 155
Helman JS 150
Herman F 31
Hilbert D 104
Hirsch P 50
Hitchcock AP 42
Howie A 50, 155
Hückel E 76
Inokuti M 23
Isaacson M 41
Jackson JD 76
Johnson D 41, 54, 161
Johnson RE 5
Jones AF 160
Jonscher AK 60
Joy DC 41, 54
Kainuma Y 155
Kittel C 65

Kloos T 153
Kohn W 77
Krivanek OL 41
Landau LD 18, 155
Lang W 65
Langkowski J 96
Langmuir I 65, 82
Leapman R 43, 118
Lehner G 79
Lifschitz EM 18
Lindhard J 78
Lopez-Rios T 88
McCarthy IE 40
Madelung O 123
Maher DM 41, 54
Manson ST 31
Maron IA 103
Massey H 9
Mathewson AG 88
Mattuck RD 98
Misell DL 154, 160
Mitroy JD 40
Moiseiwitsch BL 26
Mott NF 9
Myers HP 88
Nicholson RB 50
Ohtsuki Y-H 155
Otto A 86
Pashley DW 50
Philipp HR 55
Pines D 123, 134
Pinsker ZG 155
Platzmann PM 20, 48
Priftis GD 21
Raether H 69, 153
Raimes S 98
Raith W 40
Ramsauer C 12. 38

Rez P 154
Rinaldi R 156
Ritsko JJ 33
Robusto PF 87
Rosa R 156
Ruppin R 92
Rutherford E 1, 9
Saunderson JL 155
Sauter F 90
Schattschneider P 160, 162
Schnatterly S 33, 148
Schneider JR 48
Schwarzschild K 8
Seitz F 79
Serneels R 155
Sexl H 8
Sexl R 8
Silcox J 54, 153
Skalicky P 162
Skillman S 31
Sölkner G 58, 153
Sommerfeld A 15, 65
Sparrow TG 42, 164
Spence JCH 54, 154, 160
Stegun IA 36
Stelbovics AT 40
Takagi S 155
Teng CH 42
Tonks L 65
Townsend PD 12, 38
Trebbia P 41
Urner-Wille M 153
Van Hove L 20
Venghaus H 55
Vosko SH 77
Vradis A 21
Vuye G 88
Walecka JD 123

Wehenkel C 56
Whelan MJ 50
Williams BG 11, 42, 50, 164
Wolf PA 20
Wong J 53
Yamamoto T 155
Zapfl M 162

Subject Index

absorption edges 15
angle resolved inelastic scattering 158
angular differential cross section 3
angular momentum 6
ATR geometry 86
attenuated total reflection 86
Auger-yield 45

band width 134, 140
bands, of plasmons 148
beam spreading 154
Bethe approximation 29
Bethe differential cross section 23, 24
Bethe ridge 25, 26
Bethe surface 26, 27
billiard balls 37
black holes 8
Born approximation 28
Bragg reflection 74
bubble 128, 129

central potential 6
centrifugal barrier 7, 32
Cerenkov radiation 94
channeling 42
channeling, of electrons 50
charge density 142
    correlation 48, 138, 141
    waves 72, 142
classical impulse approximation 8
classical mechanics, deviations from 38
classical scattering 1
Clebsch-Gordon coefficients 31
collection angle 45

collection efficiency 44, 45
collective mode 78
collective oscillation 82
Compton line 11
Compton profile 48
Compton scattering 41, 47, 48
conduction band 140
   width 134, 140
conduction electrons 123, 149
conservation
   of angular momentum 38
   of energy and momentum 1, 10, 25, 48
   of energy 2, 150
   of particles 4, 129
constants of motion 6
convolution 158
correlation
   energy 132
   charge density 48, 138, 141
Coulomb gauge 63
Coulomb field 124
Coulomb force 78
Coulomb interaction 78, 100, 119, 120, 137, 139, 150
Coulomb potential 1, 18, 135, 149
Coulomb scattering 23
cross section
   angular differential 3
   Bethe differential 23, 24
   differential 16, 53, 92
   diffusion 13
   doubly differential 10
   elastic 1
   energy transfer 11
   ensemble scattering 48
   inelastic scattering 10, 57, 141
   of a hard sphere 6
   of the rare gases 12
   Rayleigh-Thompson 21

Rutherford 9, 23
  scattering 22, 56, 141
  singular 9
  total elastic 34
current density, quantum mechanical 17
cut-off wave number 148

Debye length 138
deflection function 4
  for a hard sphere 5
delayed edges 33
density
  correlation 20
  fluctuations 80
  of states 53, 54
  operator 19, 111
detection of single atoms 47
dielectric constant 60, 148
dielectric function 55, 60, 67, 88, 96, 138, 141, 145
dielectric susceptibility 60
dielectric tensor 146, 148
differential cross section 16, 53, 92
differential scattering probability 94
diffusion cross section 13
dipole moment 68
dispersion 74, 118, 131, 140
  of the collective oscillation 71
  of the free particle 125
  of the phonon 88
  of the plasmon 80, 82, 91, 148, 152, 153
  of the surface plasmon 86, 87
  of the transverse fields 72
dissipation 92
doubly differential cross section 10
dressed particle 136
drift experiments 13
Drude Model 68, 71, 139

dynamic form factor 18, 19
dynamical structure factor 141
Dyson Equation 105, 118

ECOS 42, 49
effective interaction 137
effective mass 134
effective potential 7, 31
elastic cross section 1
elastic interaction 2
elastic small angle scattering 161
electron Compton scattering 42
electron-hole continuum 127, 141, 148
ELNES 43, 53
ELS 15, 97
energy loss 10, 11, 19
    function 56, 69, 71, 95, 96, 152
    spectra, in transmission 161
    spectra 96, 155
energy transfer cross section 11
ensemble scattering cross section 48
EXAFS 42
EXELFS 42, 52, 53
external perturbations 55

fast collisions 11, 18, 23
Fermi energy 82, 124, 131
Fermi sphere 127
Fermi surface 77, 140
Fermi vacuum 112, 113, 133
Fermi vector 112
Feynman graphs 100, 121, 122, 135
fluctuation-dissipation theorem 21
fluorescent yield 46
forward scattering 96, 128
Fourier transform 58, 99, 106, 159
    of the Compton profile 48

free particle dispersion 125
frequency dispersion law 66, 118, 131, 140
Fresnel equations 83, 89
Friedel oscillation 77, 138

generalized oscillator strength 22
glory singularity 9
golden rule 16
GOS 22
gravitation 8
Green function 104, 150
    in second quantization 112
    two-particle 142
Green operator 103, 106
ground state energy 131, 134
group velocity 71, 72

hard sphere 5, 6, 36
Hartree approximation 127
Hartree equations 130
Hartree-Fock approximation 130
head-on elastic collision 1
Heisenberg operator 20
holes 112, 121
hydrogenic model 26, 44

impact parameter 4, 6, 7
impulse approximation 8
inelastic interaction 10
    angle resolved 158
    classical 17
    cross section 10, 57, 141
    form factor 19
    of photons 21
interaction
    ion-electron 149, 150
    line 142

   long range 134
   potential 17
inversion problem 156
ion-electron interaction 149, 150
ionisation energy 11

jellium 123, 149

Kramers-Kronig analysis 56, 57, 96, 164

Legendre polynomials 30, 159
Lennard-Jones potential 7, 9
Lindhard permittivity 138, 141, 152
Lippman-Schwinger Equation 101, 102, 105
local fields 146, 148
long range interaction 134
longitudinal oscillations 78
longitudinal solutions, to the Maxwell Equations 64
loss function 56, 69, 71, 95, 96, 152

macroscopic dielectric constant 148
matrix approach, for deconvolution 161
Maxwell Equations 62, 64, 65
Maxwell theory 55, 56
mean free path 4, 157, 163
Mickey Mouse 110
microanalysis 41, 44
minimum detectable mass 45
minimum mass fraction 47
Monte Carlo simulation 155
multiple plasmon excitation 153, 164
multiple scattering 153

nonlocal optics 91

oscillator strength 22

particle
  density 113
  orbits 8
  trapping 7
periodic potential 116
permittivity 21, 55, 60, 143
perturbation
  expansion 151
  theory 16, 132
  screening of the 78
phase shifts 35, 38
phonon dispersion 88
plasma frequency 68, 81
plasma 79
plasmon 73, 78, 79, 139, 141
  branches 148
  dispersion 80, 91, 148, 153
  excitation, multiple 153, 164
  polariton 73
polariton 73
polarizability 137, 143
polarization 142, 146, 149
potential
  central 6
  Coulomb 1, 18, 135, 149
  effective 7, 31
  Lennard-Jones 7, 9
  periodic 116
  pseudo- 151
propagator 100, 108, 113, 114, 117
pseudopotential 151

quantum mechanical current density 17
quasi-particles 78, 138

radial momentum 6
rainbow singularity 9
Ramsauer-Townsend effect 12, 38
random phase approximation 135, 137, 149
rare gases 12, 38
Rayleigh-Thompson cross section 21
relativity 8
relaxation time 70
resolvent kernel 109
ring diagrams 137
RPA 137
Rutherford cross section 9, 23

scattering 100, 115, 116
   angle 158
   cross section 22, 56, 141
   probability 142, 143
screening 75, 78
   length 76, 77, 138
   of the Coulomb potential 77
second quantization 111
selective summation 118, 135, 136
self consistent field 31, 129
self-energy 136, 140
shift theorem of spherical harmonics 159
short range interaction 134
site-specific excitations 50
small angle scattering, elastic 161
solid state effects 144
Sommerfeld model 125
spherical harmonics 31, 159
surface oscillation 85
surface plasmon 86, 87, 88
   dispersion 86
   polariton 85

Thomas-Fermi screening length 77, 138
time evolution operator 108

total cross section 28, 32, 34, 44
  elastic 34
total reflection 84, 87
  attenuated 86
translational invariance 145
transmission energy loss spectra 161
transverse gauge 63
two-particle Green function 142

uncertainty principle 14

virtual processes 137

width of the conduction band 131, 134, 140
Wigner 3j-symbols 31

Electron Impact Ionization

Edited by
Professor Dr. **T. D. Märk,**
Institut für Experimentalphysik,
Universität Innsbruck, Austria, and

Professor Dr. **G. H. Dunn,**
Joint Institute for Laboratory Astrophysics,
National Bureau of Standards
and University of Colorado, Boulder,
Colorado, USA

1985. 156 figures. XI, 383 pages.
ISBN 3-211-81778-6

This book is the first comprehensive review on all aspects of the kinetics and dynamics of electron impact ionization. It deals with quantum mechanical, semiclassical, and semiempirical methods of calculating ionization cross sections, and with experimental and theoretical aspects of their threshold behavior. A summary of today's knowledge about differential, partial, innershell, and total ionization cross sections is presented. Furthermore the recent subject of electron-ion ionization and, finally, the applications of quantitative knowledge of the electron ionization process in various fields such as mass spectrometry, plasma diagnostics, astrophysics, fusion, aeronomy, gaseous electronics, and radiation physics are discussed.
The book will be a major source of information for researchers and graduate students working in the field of the ionization process and its quantitative description.

Springer-Verlag Wien New York